Praise for *The Politics of Air Pollution*

"Gonzalez ... convincingly demonstrates that economic elites played a major and generally the determining role in setting the agenda for these early efforts to control air pollution."
— *Political Studies Review*

"Due to its accessibility and contributions to an understanding of environmental policy in a federalist context, this book is recommended for all readers."
— *Perspectives on Politics*

"...Gonzalez's beautifully written volume introduces readers to an urban history that they would not find elsewhere."
— *Environmental Conservation*

"...a must-read for any political scientist who studies environmental politics."
— *Environmental Politics*

"...*The Politics of Air Pollution* ... challenges several prevailing notions regarding the promulgation of air pollution and environmental policy ... Unlike the command and control modes of the 1970s, Gonzalez successfully illustrates how regional growth coalitions seek technology-based solutions."
— *American Politics*

The Politics of Air Pollution

SUNY SERIES IN GLOBAL ENVIRONMENTAL POLICY

Uday Desai, *editor*

THE POLITICS OF AIR POLLUTION

Urban Growth, Ecological Modernization, and Symbolic Inclusion

GEORGE A. GONZALEZ

STATE UNIVERSITY OF NEW YORK PRESS

Published by
STATE UNIVERSITY OF NEW YORK PRESS
ALBANY

© 2005 State University of New York

For information, contact
State University of New York Press, Albany, NY
www.sunypress.edu

Production, Laurie Searl
Marketing, Michael Campochiaro

Library of Congress Cataloging-in-Publication Data

Gonzalez, George A., 1969–
    The politics of air pollution : urban growth, ecological modernization, and symbolic
inclusion / George A. Gonzalez.
        p. cm. — (SUNY series in global environmental policy)
    Includes bibliographical references and index.
    ISBN 978-0-7914-6335-2 (alk. paper)—978-0-7914-6336-9 (pb alk. paper)
        1. Air—Pollution—Economic aspects—United States. 2. Air—Pollution—Political
aspects—United States. I. Title. II. Series.

HC110.A4G66 2005
363.739'2'0973—dc22

                                                                                2004045257

10   9   8   7   6   5   4   3   2   1

Contents

Acknowledgments

A number of individuals have read and commented on different portions of this book. Their help has significantly contributed to the strengthening of this work. I have personally and profusely thanked them all. Nevertheless, I want to take this opportunity to thank my colleagues here at the University of Miami who treated most of the chapters in this text through the American Politics Research Committee. I also want to especially thank Sheldon Kamieniecki, because, despite the fact that I graduated a number of years ago from the doctoral program in the Department of Political Science at the University of Southern California, he remains a mentor and friend. Finally, I want to give my sincerest thanks to those that gave me interviews in connection with this project. The manner in which they freely gave of their time and insights demonstrates their commitment to the advancement of human knowledge.

I also want to make special mention of my wife Ileana, my stepson Roman, my newborn daughter Alana, my fictive brother Frank Janeczek, and my mother, father, and sister. I love them all, and their love and support mean everything to me. This book is dedicated to Roman, who at six embodies the generous spirit and warmth I hope will shape our future.

The following journals have granted permission to include in this manuscript material adapted from my previously published articles:

"Democratic Ethics and Ecological Modernization: The Formulation of California's Automobile Emission Standards." *Public Integrity*, vol. 3, no. 4 (Fall 2001): 325–344. Copyright © 2001 by American Society for Public Administration. From PUBLIC INTEGRITY, vol. 3, no. 4 (Fall 2001): 325–344. Reprinted with permission from M. E. Sharpe, Inc.

"Local Growth Coalitions and Air Pollution Controls: The Ecological Modernization of the U.S. in Historical Perspective." *Environmental Politics* vol. 11, no. 3 (Fall 2002): 121–144.

"Urban Growth and the Politics of Air Pollution: The Establishment of California's Automobile Emission Standards." *Polity* 35, no. 2 (winter 2002/2003): 213–236.

ONE

Local Growth Coalitions, Environmental Groups, and Air Pollution

IT HAS BEEN over thirty years since the United States federal government enacted sweeping legislation—the Clean Air Act of 1970—to address the acute air pollution that was facing numerous urban areas (Jones 1975). Air pollution emissions nevertheless continue to persist at high levels, with several U.S. urban regions facing seemingly intractable poor air quality (Cherni 2002; Davis 2002; Lee 2004; Hebert 2004). Moreover, global warming has solidified into an accepted scientific fact (Revkin 2001; 2002 June 3; Trenberth 2001). The recent global warming trend is in large part the result of human-made airborne emissions of such gasses as carbon dioxide and nitrogen oxide (Christianson 1999; Firor and Jacobsen 2002; Jacobson 2002). Despite this scientific consensus, and the ominous signs pointing to a rapid heating of the globe—e.g., the ongoing melting of the polar ice caps (Revkin 2002 March 20)—U.S. policymakers have not enacted policies to directly abate the emission of human-made greenhouse gasses (Brown 2002; Revkin 2002 Feb. 15). Why, despite strongly worded regulatory legislation and the ample scientific data demonstrating the negative environmental and health effects of air pollution, does the United States continue to experience poor air quality in many urban areas as well as significantly contribute to global warming?[1] A primary reason the United States has not taken decisive action against airborne emissions is because the political energy to abate air pollution in the United States does not flow from a specific effort to protect human health or the environment but from a historic effort to realize wealth from the ownership and sale of land.

1

David Ricardo (1830) describes the political economy of capitalism as comprising an effort between land owners, workers, and capitalists to capture the economic benefits derived from modern production techniques (also see Foley 2003). The historic conflict over air pollution in the United States is largely the result of landed interests seeking to maximize the value of their land by minimizing the economic harm from airborne pollution. Therefore, as industrial capitalists, in the process of producing and selling commodities to capture profit, emit air pollution, landed elites seek to abate localized air pollution to capture rent from the utilization of their land. Segments of the working class have only been recently mobilized on the issue of air pollution, but this recent mobilization has not significantly altered the terms of the air pollution debate set by large land owners and industrial capitalists. This debate is centered around the deployment of technology to abate localized air pollution (i.e., the "ecological modernization" of production and transportation facilitates).

To analyze the politics of air pollution in the United States, I have deduced from the above theoretical framework a thesis with four interrelated components. Each component of this thesis is original and controversial in its own way:

1. The first component of this thesis is that U.S. air pollution abatement policies are driven by landed interests. These interests take the contemporary form of local growth coalitions—composed mostly of large land holders, land developers, and the owners of regional media and utility firms. Local growth coalitions economically profit from economic growth in particular localities (Logan and Molotch 1987).

My position on the political impetus underlying U.S. clean air policies is not shared by other scholars who analyze clean air politics. They generally hold that U.S. air pollution abatement policies flow from middle-class concerns over the negative aesthetic and health effects of air pollution. In other words, the politics over air pollution is set by most contemporary thinkers within a framework pitting relatively privileged segments of labor (i.e., the middle class) against capital (e.g., Inglehart 1977; Hays 1987; 2000; Stradling 1999).

2. The second part of my thesis is that clean air policies are functional to the operation of the market and to the realization of profit. Political scientists and historians who study U.S. pollution abatement policies have failed to realize this, with some exception (e.g., Dewey 2000), because they exclusively focus on the economic costs absorbed by industrial manufacturers as a result of clean air policies. They fail, however, to consider the economic benefits derived from cleaner air by real estate interests and other profit-driven concerns whose markets are place bound (e.g., regional media outlets and utilities). To the extent that clean air policies reduce air pollu-

tion, such policies contribute to a positive local investment climate, and, in turn, they help locally oriented economic interests realize the profits associated with such a climate.

3. The center of policymaking in the area of clean air is the urban milieu. Researchers who study U.S. environmental politics normally assume that it is the politics and policies at the federal level that are driving events in the environmental policy arena. This is particularly the case with those environmental policies instituted after the federal environmental legislation of the early 1970s (Rosenbaum 1998; Andrews 1999; Graham 2000; Kraft 2001; 2002). Even those political scientists that analyze clean air policymaking on the state and local levels (e.g., Kamieniecki and Farrell 1991; Lowry 1992; Ringquist 1993; Potoski 2001), hold that the federal government is the dominant institutional force in this policy area. Instead, however, as different local growth coalitions have moved politically to mitigate the adverse economic effects of air pollution, it is local and state governments that have taken the political and policy lead on the issue of air pollution. In light of state and local government assertiveness on the issue of air pollution abatement, the role of the federal government has in large part been to provide uniformity in the nation's clean air regime. This is most readily apparent in the politics surrounding the 1990 Clean Air Act and its content.

4. Environmental groups have been symbolically included in the clean air policymaking process. Historians and political scientists either directly argue or assume that environmental groups have a direct and positive impact on the creation and strengthening of air pollution abatement policies at all levels of government. As I will demonstrate, however, contemporary and historical environmental groups have little to do with the current approach to air pollution abatement. Moreover, it is an open question whether environmental groups help determine the present level of regulation assessed on business and industry. Their most tangible contribution to the policymaking process is to provide it with legitimacy.

After expounding on the four facets of my thesis, I conclude this chapter with a detailed overview of the book.

ECONOMIC ELITE THEORY
AND LOCAL GROWTH COALITIONS

The first component of my argument is consistent with an approach referred to as economic elite theory (Lamare 1993; 2000; Gonzalez 1998; 2001a; 2001b). Advocates of this view hold that the nation's economic elite are the dominant political force in U.S. society (Miliband 1969; Barrow 1993, chap. 1; Domhoff 2002). Clyde Barrow (1993) points out that "typically, members of the capitalist class [or the economic elite] are identified as those persons who manage [major] corporations and/or own those corporations." He adds that this group composes no more than 0.5 to 1.0 percent of the total U.S. population (17).[2] This group as a whole is the upper class and the upper echelon of the corporate or business community. The resource that members of the economic elite possess that allows them to exercise a high level of influence over the state is wealth. The wealth and income of the economic elite allow them to accumulate superior amounts of other valuable resources, such as campaign finance, and legal and scientific expertise (Barrow 1993, 16).

My focus on clean air policies, however, prompts me to forward a more refined description of the U.S. economic elite than offered by Barrow. Namely, while the strong majority of the American capitalist class derives its dominant economic, social, and political position from its ownership and control of the means of production and distribution (i.e., capital), there is a significant minority of the capitalist class that derives its wealth and status from the ownership and control of place (i.e., land)(Harvey 1985). Moreover, there are capitalists who, because of the nature of their business, are economically tied to specific locations. These types of businesses are regional media outlets, utilities, banks, real estate agencies, and law firms that are involved in real estate transactions (Molotch 1979; Bowles et al. 1983).

Therefore, the leadership and ownership of most industrial sectors, such as automobile manufacturers (Luger 2000), adopt a largely national and/or international political and policy perspective due to the nature of their economic interests. In contrast, members of local growth coalitions are forced, also due to their economic interests, to adopt an emphasis and outlook that focuses on local political and policy issues. It is because their economic well-being is tied to specific land that members of any given growth coalition are prompted to consider the issue of air quality as it relates to those locations where their economic future is vested.

My central contention is that clean air policies in the United States can only be fully understood by identifying local growth coalitions as dominant political actors who seek to ameliorate air pollution to protect local land values and local investment climates. This, however, is only part of the politics that surround the issue of air pollution. For while local growth coalitions are key actors in addressing the U.S. air quality problem, they are also key actors in its creation. A central reason why air pollution has historically been a problem in the United States is because local growth coalitions seek to attract virtually unlimited amounts of capital to an area—normally regardless of the costs. Moreover, these coalitions have historically promoted the automobile as the primary mode of transportation in urban regions. This form of transportation facilitates large land owners' and land developers' quest for profit (chapter 4 of this book). By conceptualizing clean air regulations as part of a broader set of physical and regulatory infrastructures designed to attract and maintain investment in an area, it becomes apparent how the U.S. air pollution problem was created and why our response to this problem has taken the specific form that it has—one centered on techno-logical controls.

THE COMPETITION FOR CAPITAL
AND CLEAN AIR REGULATIONS

It is a widely accepted axiom among those who study urban politics that cities, broadly speaking, can be most aptly regarded as "growth machines" (Mol-lenkopf 1983; Logan and Molotch 1987; Jonas and Wilson 1999). In other words, what virtually all students in this sub-field of political science acknowl-edge is that the desire for investment is a central political feature for almost every locality (e.g., Smith 2001; Savitch and Kantor 2002; Sellers 2002).

Given the importance assigned to local investment, state and local gov-ernments are not passive with regard to the potential location of capital investment. Instead, these government entities provide certain economic and political enticements to attract such investment. Schools, roads, and air-ports, for example, can be seen as overall efforts to attract capital to an area. In the most abstract sense, these government provided amenities are subsi-dized inputs into the production process (Gough 2000; O'Connor 2002). Additionally, state and local governments will disseminate the symbols and rhetoric necessary to assure potential investors that their investments are welcome and will be afforded political priority (Pred 1980; Eisinger 1988; Savitch and Kantor 2002).

Within this context of competing for capital investment, Paul Peterson (1981) contends that certain politics and policies cannot be engaged in by localities. He argues that locally financed and implemented income re-dis-tributive policies serve as an economic disincentive to investment in a local-ity, and this is why localities do not generally engage in such policies. More-over, income re-distribution policies create political uncertainty in a locality for investors. The concern is that such policies might lead to the excessive taxation of their investment and/or profits.

Hence, local and state governments are generally configured to provide the economic and political factors necessary to entice capital to an area. By providing these factors, these governments are helping localities to capture the economic benefits associated with such investments, including local job creation, rising land values, a larger consumer base for firms vested in the local consumer market, and a larger tax base for local government. While the boosters of local growth regularly point to the economic benefits of such growth, it is also accompanied by negative consequences for localities, such as congestion; air, water, and waste pollution; aesthetic blight; rising housing costs; and additional amounts spent on the delivery of public services (Logan and Molotch 1987; Dreier et al. 2001; Gainsborough 2001; 2002; 2003).

By taking into account both the positive and negative effects associated with local growth, we can understand the central axis of U.S. urban politics. There are politics over who is going to attain the benefits of such growth. Economically, the benefits of local growth accrue largely to the local growth coalition in the form of higher profits derived from increasing land values and an expanding consumer base. While Peterson, no doubt, provides part of the explanation as to why state and local governments do not re-distribute the economic benefits of local growth, the other part of the explanation is the ability of members of the local growth coalition to ensure that such governments are primarily configured to attract local investments and largely incapable of re-distributing the economic benefits of local growth (Stone 1989).

With regard to the job growth that accompanies increased local economic growth, Harvey Molotch shows in his landmark study that local rates of economic growth tend to have little effect on the unemployment rates of localities (Molotch 1976). What proponents of local economic growth regularly overlook when they champion local growth is that labor, like capital, is mobile in the modern era (Hernandez 2002). Thus, new jobs created as a result of new investment are just as likely to go to new residents as to natives. Significantly, new investment in a locality is predicated on this mobility and a sufficiently large and educated local labor supply (Barrow 1998; Thurow 2001; Reich 2002).

The negative effects of growth have been an increasing wellspring of political controversies in the urban milieu over the last twenty-five years. Richard DeLeon (1992), for example, outlines how in San Francisco during the late 1970s and into the mid-1980s an electoral coalition was mobilized around preventing growth in the city and the negative consequences of growth. In general, neighborhood groups throughout the country have organized politically to prevent forms of investment viewed as diminishing the quality of neighborhood life (Szasz 1994; Logan 1995; Gould et al. 1996; Ferman 1996; Tesh 2000). In addition, the poor, as well as ethnic minorities, have in certain instances mobilized against the environmental hazards in or near their residential areas (Bullard 1990; Pulido 1996; Schlosberg 1999; Cole and Foster 2001; Rhodes 2003).

Local and state governments have not necessarily been passive in the face of the negatives associated with growth. Certain localities, for example, have rejected proposed investment projects if such projects are viewed as creating too many negative externalities. Such rejections have generally been made by wealthier areas that perceive certain types of investment as inconsistent with the quality of life and high property values in those areas (Warner and Molotch 2000).

State and local governments—to varying degrees—have historically addressed the environmental hazards brought on by the manufacturing and transportation processes associated with capital investment and economic growth (Tarr 1996; Casner 1999). Statistical analyses have established that in the contemporary era it is those states and localities with the highest levels of economic activity that tend to have the highest level of regulations on water and air pollution as well as on hazardous waste (Game 1979; Williams and Matheny 1984; Lowry 1992; Ringquist 1993; Davis and Davis 1999; Potoski 2001).

With this co-existence between high levels of economic activity and comparatively high levels of environmental regulations, these factors are obviously compatible. This compatibility results from the fact that environmental regulations on pollution rely on technology to abate pollutants (Lowry 1992; Ringquist 1993; Grant 1996; chapters 3, 5, and 6 of this book).

Approaches that rely on technology to abate pollution are compatible with local economic activity and growth in two key ways. First, such approaches do not seek to curb local economic activity or growth to affect lower levels of pollution. This would be the most assured means to reduce pollution. Second, a reliance on technology to abate pollution is politically reassuring to industrial firms. This is because it is these firms that control the development and deployment of such technology. Thus, any policies that rely on pollution abatement technologies rely on industrial firms to develop and deploy this technology. In this way, industrial firms are central in the making and implementation of those policies that utilize pollution abatement technologies (Noble 1977; Davison 2001; Hornborg 2001). For such firms, a reliance on technology to abate pollution is a much more politically palatable approach than one, for example, that would seek to dictate production schedules.

Local Growth and Technological Controls on Air Pollution

The question for researchers is what factors have resulted in policy approaches that rely on technology to reduce air pollution—the specific focus of this study? For while industrial firms are not that adversely affected by such policies, they have no specific interest in initiating air pollution abatement policies, even those that rely upon technology. Researchers generally argue that technology-based air pollution abatement policies are in large part the result of interest group competition (Bryner 1995; Graham 2000). In this competition, industrial firms are on one side, and environmental groups are on the other side (Marzotto et al. 2000). This competition leads to a compromise on technological controls to manage pollution. In this way, air pollution emissions are addressed, which pleases environmental groups and an environmentally minded public, but the imperatives of economic growth and industrial firms' control over production remain largely unaffected. The latter are the primary political goals of industrial producers. Sabatier (1987) holds that this compromise is internalized by these competing interests in a form of "policy learning." Public officials, both elected and nonelected, also embrace the technological approach to address air pollution as a way to reconcile the public's desire for economic growth, automobile usage, and cleaner air (Dryzek 1996a, 480; Rajan 1996; Mazmanian 1999).

Maarten Hajer (1995) holds that the current political emphasis on technology to address such environmental bads as air pollution is the result of an autonomous society-wide discourse, known as "ecological modernization." Inherent in this discourse is the notion that through "clean" technologies economic activity and growth can and should have a benign relationship to the environment (also see Weale 1992 and Bernstein 2001).

Technological controls on air pollution, however, are precisely in line with the economic and political interests of local growth coalitions. This is the case for three interconnected reasons. First, as noted above, cleaner air results in an improved local investment climate. Second, government mandated technological controls affect the output of air pollution, but they do not regulate the rates of local growth. This allows local growth coalitions to freely promote and embrace increasing amounts of local investment and population growth. Finally, the promotion of technology by local growth coalitions to address air pollution maintains—for reasons already discussed—investor political confidence in a specific geographic area.

Therefore, I challenge the view that technological controls on air pollution result from a compromise between competing interest groups, or that this approach to air pollution abatement results from public officials seeking to reconcile the rather conflicting desires of the public. Moreover, as I will demonstrate in the subsequent chapters of this study, the efforts to manage air pollution through technology significantly predates the development and dissemination of the ecological modernization discourse (chapters 3 and 5). Next, I discuss how the decentralized nature of the U.S. clean air regulatory regime allows local air quality and clean air policies to become factors in the competition between localities for capital investment.

THE FEDERAL GOVERNMENT
AND AIR POLLUTION ABATEMENT

The federal government is a secondary actor in the formulation and implementation of the United States's clean air regime. It only began to take a substantive policy role on the issue of air pollution with the passage of the Clean Air Act of 1970. Prior to this legislation, the federal government limited its activity largely to the funding of scientific research (Rosenbaum 1998; Andrews 1999). Therefore, while cities in the Midwest, like Chicago and St. Louis, were inundated with smoke during the late nineteenth and early twentieth centuries from the burning of coal, Washington, D.C. was largely idle on this matter.

In this period, and leading into the 1970s, the substantive politics and policies dealing with air pollution were taking place in various localities. It was those areas in particular that had a heavy dependence on coal that faced the most significant air pollution problem. Moreover, it was those areas that were most dependent on soft coal where air pollution was most acute and persistent. Not surprisingly, it is in these areas where the politics of air pollution were most intense. Major cities that were plagued with extremely poor air quality, and with the highest level of political activity surrounding the issue of air quality, included New York, Pittsburgh, Chicago, St. Louis, Cincinnati, and Birmingham (Grinder 1980; Stradling 1999; Dewey 2000).

Once industrial manufacturers, utilities, and railroads in the post–World War II period moved in large part to oil and natural gas as a source of fuel, cities like Chicago and Pittsburgh experienced drastically improved air quality. During this period, however, a new threat to clean air emerged—the automobile. Now cities such as Los Angeles began to face serious air pollution problems. It was the air pollution politics arising from Los Angeles that led to the first serious effort to address airborne pollution from the automobile. Again, the federal government remained largely on the sidelines in the 1950s and 1960s while states like California undertook major efforts to curb automobile emissions (Krier and Ursin 1977).

It was only after the environmental social movement of the late 1960s and early 1970s that the federal government assumed a direct role on the issue of air pollution. The federal government's post-1970s effort in this area, however, has been criticized as weak. Matthew Cahn (1995), drawing inspiration from the work of Murray Edelman (1964; 1988), goes so far as to label federal air pollution policies as "symbolic." He holds that they serve more as a mechanism to assuage public opinion than a concerted attempt to remedy the nation's air quality problems. Indicative of the weak federal response to air pollution is the fact that leading into 1990, the federal agency responsible for regulating air pollution emissions, the Environmental Protection Agency (EPA), had only set regulations for seven hazardous or toxic airborne emissions. The Clean Air Act of 1990 was designed to rectify this. It specifies that the EPA must set standards for 189 hazardous or toxic chemicals (Bryner 1995). In this, as well as in other pollution abatement efforts, however, the EPA is hampered by an inadequate budget (Yeager 1991; Mintz 1995; Rosenbaum 1998; Weber 1998; Seelye 2003). Moreover, in the central area of automobile emissions, the federal government only took a direct role in regulating these emissions in 1980. Additionally, federal automobile emission standards have always lagged behind California's, and they currently are not as strong as those in major states as New York and Massachusetts.

On the issue of policy implementation, the federal government relies primarily on state and local agencies to formulate clean air implementation strategies. As set under the terms of the Federal Clean Air Act of 1970, states must submit State Implementation Plans (SIPs) to the EPA for its approval. These plans must outline how the individual states are going to meet federal clean air standards. The individual states are offered discretion in certain areas to develop standards that exceed federally mandated standards.

More significant is the fact that states are responsible for implementing or enforcing air pollution regulations (Game 1979; Wood 1988; 1992; Scheberle 1997; Morag-Levine 2003). States, however, have varied records concerning the enforcement of their clean air regulations. While some states may be aggressive on enforcement issues, others have been found to be very lax—even failing to enforce the minimum federal standards (Game 1979; Cushman 1998 June 7; Seelye 2001; "Lung Association" 2002).

Thus, the reliance on states to formulate and implement air pollution abatement programs has lead to an uneven clean air regulatory regime. Defenders of this approach hold that it does have its benefits. Most specifically, it is averred that the use of state agencies to formulate and enforce the clean air regulatory regime allows government regulators to be sensitive to state and local conditions, and hence it leads to a more efficient and effective policy regime (Nice 1987). Little or no empirical evidence, however, has been generated to substantiate this oft-repeated claim.

Much more likely is that state and local officials guard their discretion over federal environmental policies (e.g., Cushman 1998 August 5) because they can calibrate environmental regulatory regimes to growth strategies. Those areas with high levels of investment and growth can use environmental policies, including clean air policies, to mitigate the environmental negative externalities of economic activity and growth, thereby seeking to insure that existing growth and investment levels are not jeopardized by such externalities. Whereas those states and localities that have comparatively lower levels of growth and investment simply may not have the political will or desire to regulate emissions, because these emissions, to the extent that they do exist in these areas, do not present a perceived threat to growth. In other words, the pollution "load" has not reached a threshold that represents a threat to the local business climate. Local growth boosters in less developed areas may also perceive pollution abatement politics and policies as a disincentive to potential local investment.

These hypotheses are supported by the aforementioned statistical analyses comparing the "strength" of air pollution regulations in the various states. Such studies have measured and compared the restrictive content of state air pollution regulations as well as expenditures toward the enforcement of these regulations (Game 1979; Kemp 1981; Lowry 1992; Ringquist 1993; Potoski 2001). What these studies have found is that the strength of clean air regulatory regimes vary positively with the level of wealth and economic activity in a state. Thus, those states with high levels of wealth and economic activity tend to have stronger air pollution regulatory regimes, whereas less wealthy and less economically active areas have weaker regimes.

With the nation's clean air regime fluctuating according to state and local concerns over economic growth and/or other locally or regionally generated political issues, what is the precise role of federal clean air policies? I have already noted one function of federal environmental laws and policies—legitimation. Again, it is Cahn (1995) who explicitly develops this position. As I alluded to above, it was in the face of the social movements of the late 1960s and early 1970s that the federal government passed its most sweeping environmental legislation, and created its environmental regulatory agency, the EPA (Rosenbaum 1998; Andrews 1999). In light of this context, it can be argued that this legislation and agency were created to demobilize those who were protesting against corporate America's treatment of the environment, thus bringing social and political peace to U.S. society (Ford 2001). Their creation communicated to this social movement that its environmental concerns were being addressed, and therefore protestors could end their disruptive activities and go back to resuming their normal lives.

Despite the strong wording of this early legislation, however, the federal government has not developed an aggressive environmental regulatory regime. Specifically, throughout most of its history, the EPA has lacked appropriate funding levels and strong political backing (Mintz 1995; Lee 1996; Weber 1998; Seelye 2002; 2003). Moreover, it relies largely on the states for enforcement purposes.

Cahn does not limit his critique to the relatively weak nature of the federal government's environmental regulatory regimes. He also points out that despite the rhetorical commitment of most federal elected officials to environmental protection, the federal government's overriding and primary goal is the promotion of economic activity and growth. Such activity and growth is the chief cause of air, water, and waste pollution. The federal government has also promoted and subsidized the use of highly polluting fossil fuels. The federal government's subsidizing and encouragement of fossil fuel usage is reflected in its 1998 appropriation of $200 billion for the maintenance and expansion of the nation's transportation infrastructure, which "amounted to the largest public-works program in the nation's history." Over 80 percent of these funds were dedicated to highway and bridge construction (Andrews 1999, 303–304; also see Hulse 2004). Therefore, while federal officials (since at least the early 1970s) have been publicly committed to environmental protection, their substantive policies toward this end are relatively weak. Conversely, they have aggressively pursued policies that lead to environmental degradation.

Apart from the legitimation and symbolic aspects of federal clean air policy, on certain air pollution issues, the federal government does provide uniformity and regulatory regime stability. This is most readily apparent in the realm of automobile and fuel emissions. The Federal Air Quality Act of 1967, for example, sought to establish one national automobile emission standard when states such as New York and California were setting their own standards (Krier and Ursin 1977, chap. 11). In addition, I (2001a, chap. 6) explain that the primary motive underlying the formulation of the Federal Clean Air Act of 1990 was to stabilize the nation's automobile and fuel emission standards.

Specifically, I point out how in the mid- to late 1980s state and local-level policies were threatening to create a multi-tiered system of regulations on automobile and fuel emissions. In the area of automobile emissions, leading into 1989, the United States had a two-tier regulatory structure, with California having one standard and the federal government having its own weaker standard. Within the span of a few months, this system was substantially complicated. Specifically, in the summer of 1989, New York, New Jersey, and the New England states announced their intention to adopt the California emission standard. In September of that year, California announced its plan to establish an even stronger set of standards than it currently had in place. This created the potential of a three-tier automobile emission standard system: (1) the existing federal standard; (2) the old California standard (adopted by New York, New Jersey, and New England); and (3) the newly announced California standard. And while these actions involved only a handful of states, they collectively composed about 20 to 30 percent of the U.S. automotive market.

Additionally, these state-level actions created a potentially significant legal and political dilemma. Up until this time, it had been both politically and legally accepted that states could either adopt the California automobile emission standard or the weaker federal standard. The possibility of a three-tier emission system created the legal and political uncertainty about whether states could adopt an emission standard that varied from the three standards that were seemingly going to come into existence.

Before the policy actions of these states took effect, thereby creating a significant production and distribution—as well as a legal and political—problem for automobile manufacturers, the federal government stepped in with the Clean Air Act of 1990. With this act, the federal government adopted the existing California automobile emission standard, and thus reinstated a two-tier automobile emission regulatory system. Moreover, erasing any legal ambiguity concerning the ability of states to establish their own automotive emission standards apart from those of the federal government's or California's, the Clean Air Act of 1990 contains a "no third vehicle" clause. Legal scholar John Ridge (1994) explains that this aspect of the Act prevents "the states from adopting new tailpipe emissions standards which cause or have the effect of causing the [automobile] manufacturers to have to create a new vehicle or engine" apart from those prompted by the federal and California emission standards (176).

In the area of fuel emissions, the federal government had no regulations until the Clean Air Act of 1990. In the mid- to late 1980s, however, states and localities did begin to look to regulations on fuel as a means to reducing air pollution from automobiles. In 1987, Colorado, for example, began a program that mandated that gasoline sold during the winter months contain oxygenated additives ("Colorado's High-Oxygen Fuel Test Runs Smoothly" 1988). The addition of oxygenated additives, such as ethanol or methyl tertiary butyl ether (MTBE), reduces the amount of smog producing chemicals created when gasoline is burned. Further, prior to 1990, programs in Washington state, California, New York City, and British Columbia began experimenting with alternative-fuel vehicles (Wald 1989).

From the perspective of the oil industry, both alternative fuels and gasoline additives are problematic solutions to automobile emissions. Alternative fuels, such as methanol, electricity, or natural gas, which can be substantially less polluting than conventional gasoline, represent a threat to the oil industry's lucrative gasoline market. Automobiles, however, have already begun to reach the limits of emissions reduction while burning conventional gasoline. By 1989 automobiles had reduced the emission of hydrocarbons by 80 percent, and nitrogen oxides by 60 percent, when compared to automobiles produced in the 1960s (U.S. Congress 1990, 227).[3]

Fuel additives, while used with gasoline, do represent a production problem for the oil industry. Because oil-based fuels such as gasoline, jet fuel, heating oil, and heavy fuel oil are all produced from the same barrel of oil and in the same production process, adding an additive to gasoline results in complications in producing other oil-based fuels. Specifically, generally half of a refined barrel of oil becomes gasoline and the other half becomes other forms of fuels, and there is very little that can be done to alter this (Lippman 1990). Hence, a reduction in the production of gasoline to make room for a substantial amount of additive will result in a corresponding reduction in other fuels. Conversely, to refine more oil to produce greater amounts of other fuels would increase the amount of gasoline beyond market demand.

The result of the Clean Air Act of 1990 in the area of fuel emissions was to preempt those state and local experiments in fuel additives and alternative fuels.[4] It calls for the sale of so-called clean gasoline in the nine most polluted cities: Los Angeles, Houston, New York, Baltimore, Chicago, Milwaukee, Philadelphia, San Diego, and most of Connecticut. Other cities can opt into the program. The legislation set gasoline emission standards for the gasoline sold in these nine areas, but the oil industry would determine how to meet these standards (Weisskopf 1990; Adler 1992, 36; Cohen 1995, 167–168).[5]

Therefore, the entrance of the federal government in the early 1970s into the air pollution abatement regime has not replaced state and localities as the predominant jurisdictions where this regime is formulated and implemented. The most significant policy impact of federal clean air rules has been to provide uniformity for automobile and gasoline producers. This uniformity has resulted in an automobile and fuel emission regime that limits costs and political uncertainties for these producers. In contrast, it has politically and legally undermined the ability of certain states and localities to force automotive and oil producers to provide cleaner automobiles and fuels.

THE ROLE OF ENVIRONMENTAL GROUPS IN THE MAKING OF CLEAN AIR POLICY

In his analysis of contemporary U.S. environmental politics, David Schlosberg (1999) asserts that "there is no such thing as environmentalism." He goes on to explain that "'environmentalism' is simply a convenience—a vague label for an amazingly diverse array of ideas that have grown around the contemplation of the relationship between human beings and their surroundings" (3). Reflective of this diversity, John Dryzek (1997) catalogs the different points of departure taken by various philosophers, researchers, and activists in conceptualizing the environment and the means to address perceived environmental problems.

While we can speak of the varied and differentiated meanings of environmentalism and even sustainability (Vos 1997; Bernstein 2001), those groups that directly lobby government, and refer to themselves as environmental groups, largely limit their advocacy to the ecological modernization of business, industry, and the automobile. Thus, as it relates to pollution issues, those environmental groups incorporated within the public policymaking process limit their advocacy to urging the government mandating of "clean" technologies. In the lobbying activity leading up to the passage of the 1990 Clean Air Act, for instance, the two central objectives of the lead environmental group in this lobbying effort, the National Clean Air Coalition (NCAC), were dependent on technology.[6] The NCAC is an umbrella organization composed of public service groups concerned with clean air. It is made up of virtually all major environmental groups, as well as church groups, civic and public health groups, and labor unions. The NCAC's key objectives were a mandatory two-round strengthening of the federal government's automobile emission standard and the mandating of the production of one million alternative fuel automobiles by 1995 (Gonzalez 2001a, chap. 6). Neither of these technologically dependent proposals made it into the final legislation.

The one aspect of the Clean Air Act of 1990 that can be attributed to an environmental group is the legislation's acid rain provision. The Environmental Defense Fund (now Environmental Defense [ED]) is credited with directly crafting that portion of the Act that provides for a permit trading system that is expected to induce the reduction of sulfur dioxide emissions from power plants in the Midwest and Appalachia. These emissions fall upon the Northeast and Canada in the form of acid rain. Under this system, polluters who reduce their sulfur dioxide emissions below a prescribed amount can sell permits or "allowances" that equal the total of their excess reduction to other power plants that are above the set limit. In turn, firms that exceed their emission limit must purchase permits from those firms that have reduced their emissions below government standards (Bryner 1997; Alm 2000; Ellerman et al. 2000, chap. 2; Gorman and Solomon 2002).

This permit-trading approach to pollution abatement can be said to be means neutral, since it only serves to internalize the costs of pollution and does not dictate how emissions are to be reduced. Nevertheless, this portion of the Clean Air Act of 1990 eschews an obvious nontechnological approach that would assuredly reduce sulfur dioxide emissions from power plants. Specifically, the Act does not prohibit the usage of high-sulfur (bituminous) coal and mandate the utilization of low-sulfur (anthracite) coal. Low-sulfur coal, a viable substitute for high-sulfur coal, is a low-tech and inexpensive means to reduce sulfur dioxide emissions from power plants (Ackerman and Hassler 1981). In lieu of such a prohibition, the acid rain permit-trading system formulated by ED can be viewed as creating an incentive structure to prompt the development and implementation of pollution reduction technology (e.g., scrubbers). At a minimum, it allows firms to utilize technology to address the environmental problem of acid rain.

State level environmental groups that lobby government also limit their lobbying activity to the advocacy of technological fixes for pollution problems (Lowry 1992; Ringquist 1993; Grant 1996; Rajan 1996). This is true even of the California groups. California is the state that, according to the aforementioned studies, had the "strongest" air pollution regulations. Here a central political goal for environmental lobbyists is the state-mandated sale of alternative fuel automobiles. Leading into the late 1990s, California's alternative automotive fuel program was the most ambitious in the country, and probably the world (Grant 1996; chapter 6 of this book).

As I have explained, however, the reliance on technology to address pollution places industry at the center of the formulation and implementation processes. This is most aptly evident with the stipulation in the federal Clean Air Act of 1977 that industrial air pollution regulatory control efforts remain consistent with the "best available technology." With industry controlling and/or financing society's research and development infrastructure, it makes such technology available. Hence, as long as federal, state, and local air pollution reduction efforts are focused on technology, these government bodies are going to be directly dependent on the willingness of private firms to invest in the research and development of pollution control technologies.

While certain political and regulatory victories can be attained for environmental groups through technology, such victories cannot change the direction of the U.S. political economy. Those seeming victories attained by environmental groups through the mandating of technological controls do not change or challenge the overriding political objectives of economic growth and profit—again, currently the primary causes of human-induced environmental damage (Gorz 1985; Dryzek 1987; Cahn 1995; Luke 1997; Davison 2001; Bednar 2003).

One obvious reason as to why environmental groups incorporated into the policymaking process accept the limiting of the debate over air pollution to one of technology is because they do not have the influence to change the agenda. Hence, industry, local growth coalitions, and/or the public is exercising the "second-face of power" to prevent environmental groups from interjecting the issues of economic growth, the location of investment, modes of transportation, or profit into the policymaking process as it relates to air pollution. In other words, these factors are preventing central or core issues from making it onto the clean air agenda (Bachrach and Baratz 1962; Lukes 1974; Lindblom 1982).

We must be sensitive to the fact, however, that not all self-proclaimed environmental groups are oppositional in orientation. Thus, not all environmental groups seek to challenge the tenets and central features of the U.S. economy. ED, for example, embraces the primary aspects of the U.S. economy, and instead makes it its key political goal to reform the U.S. economy at the margins. ED is a leader in what Mark Dowie (1995, chap. 5) calls "third wave environmentalism." He views the advocates and adherents of this movement as "conservative" environmentalists because key in their thinking:

> Is the notion that production decisions should remain in the private sector and that removing market barriers and government subsidies that promote environmentally unsound practices will allow the mechanisms of the marketplace to motivate industries to make environmental protection profitable. (108)

Dowie (1995) goes on to argue that "another implicit tenet of the third-wave ideology is that all non-fraudulent businesses and industries deserve to exist, even if their technologies or products are irreversibly degrading to the environment" (108).

Regardless of the motivations of incorporated environmental groups in advocating technological fixes for environmental problems, the most tangible accomplishment of the inclusion of environmental groups is to communicate to the public that the policymaking process is inclusive, permeable, and democratic. Hence, the inclusion of environmental groups enhances the legitimacy of the state and the policies utilized to address air pollution.

OVERVIEW OF THE BOOK

The central component of my thesis is that locally oriented economic elites provide the key political capital underlying the political efforts to ecologically modernize business, industry, and the automobile. They do so to limit the negative externalities generated from local economic growth and the reliance on the automobile as a local mode of transportation. The negative externality of air pollution serves as a threat to the business climate in localities with high levels of economic activity. With local growth coalitions serving as the central political agents in the development of the nation's clean air regime, local and state governments are the central institutions in the formulation and implementation of clean air policies. There is some exception to this in the area of automobile and fuel emission regulations. The federal government, however, has taken an important role on these issues only because actions on the state and local levels were legally and economically threatening to automobile and gasoline producers. Finally, with local growth coalitions and industrial firms economically and/or politically benefitting from technology as the means to address air pollution, environmental groups have been incorporated into the policymaking process on the basis of this technology-based agenda.

It is not enough, however, to explain that the control of air pollution through technology is in line with the economic and political interests of local growth coalitions and, to a lesser extent, industry. Political scientists must connect objective interests with the state (abstractly speaking) and the policymaking process. Simply because a policy economically and politically benefits a certain group does not mean that that group actually determined the outcome in question. (Although it is one indicator of influence.) Instead, a more specific effort must demonstrate how such a group conceptualizes its interests and then acts to forward said interests, and to what extent is it able to do so (Dahl 1958). Hence, it is my position that any serious student of state behavior must determine who decides as well as who benefits? In the next chapter, I describe competing policymaking models, which outline theoretically how groups and individuals—including officials within the state—formulate policy proposals and have those proposals incorporated within public policy (chapter 2). These models will move us a step closer to understanding who benefits, who decides, and why, with regard to the U.S. clean air regime.

Chapter 3 describes the politics of air pollution during the late nineteenth and early twentieth centuries. This period is significant because it is during this era that air pollution first enveloped U.S. urban centers. In this chapter, I identify the political energy historically underlying air pollution abatement efforts in the United States. This energy emanates from leading members of local growth coalitions seeking to abate air pollution through technological controls in order to improve and/or safeguard the local business climate. Because of the costs associated with technological controls on air pollution and their unreliability during this period, these early efforts to manage air pollution and improve local air quality failed. In chapter 4, I outline how U.S. real estate interests during the 1920s successfully promoted the automobile as the primary means of transportation in the urban realm. They did so because the automobile was and is an effective means to increase the utility, and, hence, the value of land. This explains why, when the automobile became the primary factor behind Los Angeles's extremely poor air quality in the post–World War II period, the local growth coalition successfully promoted technological controls on automotive emissions and not the restriction/supplanting of automobiles in the Los Angeles basin (chapter 5).

In chapter 6, I analyze the role of environmental groups in the policymaking process as it relates to air pollution. I hold that with contemporary local growth coalitions advocating technological controls to address air pollution, and this approach dominating the discussion within government, environmental groups should withdraw from the policymaking process. Environmental groups' most tangible contribution to the policymaking process is to enhance its legitimacy and help build public support for clean air policies that leave many of our urban areas inundated with air pollution and allow the United States to be the primary global source of greenhouse gasses. Instead of allowing themselves to be symbolically included within the policymaking process, those environmental groups interested in effectively addressing air pollution should redirect their energies and resources toward educating the public on how our clean air policies are failing us. In the conclusion of the book, I bring together the argumentation articulated in chapters 1 and 2 with an overview of the empirical information posited in chapters 3, 4, 5, and 6. This allows me to draw precise conclusions, and speculate about the future of U.S. clean air policies, with specific attention to policies that relate to climate change.

TWO

Political Economy and the Policymaking Process

IN THE PRECEDING CHAPTER, I described the structural relationship between local growth coalitions, industrial firms, air pollution, and the technological controls on such pollution. As I have already outlined, local air pollution is an economic negative for local growth coalitions, whose members rely on local economic growth for profit and their overall economic well-being. From the perspective of these coalitions, technological controls on air pollution are the optimal solution to air pollution. This is because such controls abate air pollution without regulating the primary cause of this pollution—local economic growth. Technological controls on air pollution are also an acceptable solution to air pollution for industrial firms, since it is they who control the development and implementation of this means of air pollution abatement.

My argument however is not limited to a structural analysis of the political economy of air pollution and pollution controls.[1] Of equal importance here is the identification of the processes that over history have politically prompted the development and deployment of technology to address air pollution. Thus, while technological controls on localized air pollution politically and economically benefit local growth coalitions, did environmental interest groups nevertheless force their utilization? Similarly, have such controls been embraced by local political elites in an attempt to manage and reconcile the conflicting and competing desires of the public for cleaner air and a robust local economy? Alternatively, are public policies that seek to utilize technology to address localized air pollution the result of the political efforts of local growth coalitions acting on behalf of their very specific interests?

In this chapter, I describe two competing models of the government policymaking process. These models allow me to analyze the political forces that have mobilized around the issue of air pollution, and how these forces have affected public policy or state behavior. The first is the state autonomy/issue networks model, and the second is economic elite theory. The state autonomy/issue networks model can be viewed as a modified version of pluralism. Its proponents hold that actors within the state, who often act autonomously, are central to policymaking in the United States.[2] In this model, public officials, in formulating policies, will often incorporate various public interest advocates into the policymaking process—including environmental groups.

The economic elite model places economic elites at the center of the policymaking process. It is their preferences that determine the deployment of the resources and organizational methods that comprise the state (Mitchell 1991; Gonzalez 1998, 295–299; 2001b). Economic elite preferences on public policy issues, in turn, are shaped by the operation of the political economy. In this regard, these elites seek to deploy regulatory policies that improve the operation of the capitalist economy and their market position.

This chapter proceeds in the following manner. First, I begin with a discussion of pluralism and its relationship to the state autonomy/issue networks model. Second, I lay out the state autonomy/issue networks model. In analyzing this model, I discuss the different forms that the incorporation of public interest advocates within the policymaking process can take (i.e., governing coalition, regime coalition, or symbolic inclusion). Third, I treat the economic elite model, and I conclude with a discussion on the relationship between economic elites, economic activity, and regulatory policies.

PLURALISM

Pluralism, in terms of the contemporary literature on U.S. politics, represents the first attempt to arrive at an analytical understanding of policy formation. Proponents of this view posit that U.S. state behavior (i.e., decision-making) is largely the result of the political behavior of interest groups and their relationship to public officials (Dahl and Lindblom 1953; Dahl 1956; 1961). Gary Bryner (1995), for example, provides a pluralist interpretation of the passage of the Clean Air Act of 1990. He specifically argues that the act was the result of competition and cooperation between different interest groups and government officials. The difficulty with Bryner's argument, and with pluralism in general, is that neither make reference to the specific operation of capitalism. In the case of Bryner's analysis, one is left wondering why it is the Clean Air Act of 1990 and not 1985, or 2001? He does suggest that the change in presidential administrations (from Reagan to Bush), as well as changes in congressional leadership, facilitated the passage of the act (114). These factors strike one as rather ad hoc. This is especially the case when one considers the fact that the Democratic Clinton administration, even with a Democratic-controlled Congress during its early tenure, made no significant moves to strengthen federal clean air laws. Moreover, and perhaps more importantly, the factors that drive interest group proposals are generally exogenous to the pluralist model. Hence, Bryner's interpretation of the Clean Air Act of 1990 provides little insight into why interest groups lobbied for or against specific proposals.

What is missing from the pluralist approach is political economy. In other words, the pluralist conception of politics and the policymaking process does not incorporate a theory of capitalist development or some concept of capitalism. An approach rooted in political economy provides greater explanatory power than an approach that almost exclusively focuses on the interaction between interest groups and state officials. Emphasizing the operation of the U.S. political economy in the late 1980s, for example, one can see how the Clean Air Act of 1990 was in response to state and local government efforts to regulate air pollution. State and local actions during the mid- to late 1980s were threatening to hamper the operation of the national economy. The objective of the industrial sectors that participated in the formulation of the Clean Air Act of 1990 was to rationalize clean air policies in the United States under one regulatory regime (Gonzalez 2001a, chap. 6).

Dryzek (1996a) argues in his critique of traditional pluralism that "irrespective of what interest groups seek, states must meet certain imperatives" within capitalist societies (478). Hence, as Dryzek explains, pluralism fails as a policymaking model because it does not orient the researcher to the operation of capitalism and how it creates certain imperatives for the state. Following neo-Marxist views of the state and politics (Poulantzas 1973; Barrow 1993, chap. 2; Aronowitz and Bratsis 2002; O'Connor 2002), Dryzek (1996b) holds that these imperatives are forwarding the private accumulation of capital (i.e., maintaining a strong economy) and maintaining the legitimacy of the state within the context of a market economy. A potentially more promising approach in understanding the role of the state within capitalist society than that offered by pluralism is the state autonomy/issue networks model. This model, like pluralism, emphasizes the role of interest group competition in the formation of public policies.

STATE OFFICIALS AND ISSUE NETWORKS

At the core of the state autonomy/issue networks model is the notion that officials within the state can and do behave autonomously of all social groups.[3] Officials within the state have special theoretical significance in relation to the political and economic instabilities created by the operation of capitalism, because they are often looked upon to deal with such instabilities. Moreover, they are also provided in many instances with the resources, such as legal authority and a budget, to do so (Poulantzas 1973; Nordlinger 1981;

Skowronek 1982; Skocpol 1985; Finegold and Skocpol 1995; Klyza 1996; Carpenter 2001; Aronowitz and Bratsis 2002; O'Connor 2002).

The second component of this model is issue networks. The concept of issue networks was pioneered by Hugh Heclo (1978), who holds that the social movements of the late 1960s and late 1970s activated numerous public interest groups, which in turn sought to influence the policymaking process. The introduction of these public interest groups into the policymaking process converted this process from predominantly a relationship between a few powerful interest groups and public officials (Schnattschneider 1960; McConnell 1966; Lowi 1979) to one of more complicated issue networks (Baumgartner and Leech 1998). Within issue networks, public interest advocates offer perspectives and political resources (e.g., electoral votes) different than those posited by more narrowly oriented interest groups, such as business trade associations (for examples, see Heinz et al. 1993).

In this context, autonomous policymakers can and do draw upon different members of any given issue network to determine how to prioritize various imperatives and how to address them (Skocpol 1992; Skocpol et al. 2000). In this way, public interest advocates are incorporated into the policymaking process. Scientists and experts have specific importance within the state autonomy/issue networks model. This is because they offer the technical know-how to instruct public officials. Scientists and experts also orient state officials to the instabilities that must be addressed in order to avoid more serious economic, political, or social difficulties (Rich 2004). According to Theda Skocpol (1986/87), the legitimacy and usefulness of experts is enhanced by the fact that they "most often . . . attempt to act as 'third-force' mediators, downplaying the role of class interests and class struggles and promoting the expansion of state or other 'public' capacities to regulate the economy and social relations" (332). Indicative of the state autonomy/issue networks model, Michael Kraft (1994) argues that it was the intersection of positively disposed political leaders, public opinion, environmental groups, and scientific studies that lead to the formulation and enactment of the 1990 Clean Air Act.

INCORPORATION WITHIN THE POLICYMAKING PROCESS

The incorporation of public interest groups and advocates within the policymaking process, however, can take different forms. Moreover, the type of incorporation has direct implications for public policy development. The forms that incorporation can take are: (1) a governing coalition (Browning et al. 1984; 1989); (2) a regime coalition (Stone 1989); or (3) symbolic inclusion (Dryzek 1996a).

The first two forms of incorporation were posited by researchers to analyze urban politics. This is appropriate, because urban governments have been leading public agencies in U.S. air pollution abatement efforts. The federal government only sought to effectively regulate air pollution emissions in the 1970s, whereas certain urban governments were seeking to control air pollution as early as the late nineteenth century. Even when the federal government did enter the regulatory fray on this issue, its efforts continued to lag behind important states and localities. Currently, for example, the federal automobile emission standards are not as strict as the California standards. Thus, as explained in the preceding chapter, U.S. clean air policies, in many important regards, can only be understood by making specific reference to urban and state politics.

THE GOVERNING COALITION AND REGIME COALITION APPROACHES

The governing coalition approach posits a generally fluid and open-ended view of urban politics, where ethnic minorities and other interest groups utilize electoral politics to become members of governing urban coalitions. At the center of these coalitions are public officials. Utilizing the example of Los Angeles, Raphael Sonenshein (1993) argues that Mayor Tom Bradley, from 1973 to 1992, was at the center of a governing coalition composed of ethnic minorities, progressive whites, and the downtown business community.

In the governing coalition approach, business elites will normally play a prominent role in any given coalition because they provide the central political resource of campaign finance. Additionally, the growth agenda of land owners, developers, and other pro-growth interests is serviced by political elites, because capital investment in an area is necessary for local job growth and a growing economy (Peterson 1981). Nevertheless, public interest groups, representing ethnic minorities and noneconomic perspectives, can have their policy agendas addressed through the successful mobilization of voters and alliances with specific political leaders.

In explaining the variation in state clean air regulations, for example, Ringquist (1993) and Potoski (2001) point to the correspondence between environmental interest group membership in the individual states and the "strength" of such regulations. The positive correlation between environmental interest group membership and the strength of state clean air regulations would suggest that those groups with relatively high membership levels have been successfully incorporated into the governing coalitions of their respective states and/or localities. This inclusion has resulted in more restrictive clean air regimes.

The regime coalition approach offers a view of urban politics where business is the central actor in any given coalition, and not political leaders. Here coalition members are incorporated into governing regimes not necessarily based on whether they can mobilize sufficient votes, but whether certain groups, and their leaders, can potentially interfere with the local economic elites' growth agenda. In light of this potential threat, the leaders of public interest groups are incorporated and group members are provided "side-payments" in order to pacify them. This allows economic elites' growth agenda to proceed more or less unabated. Clarence Stone (1989) explains how African Americans were incorporated into Atlanta's regime coalition during the Civil Rights Era in order to maintain a positive business climate. In his analysis of San Francisco politics from the mid-1970s to the early 1990s, DeLeon (1992) describes one of the few examples of successful public mobilization efforts to obstruct local economic growth. This broad based anti-growth movement, allied with specific political leaders, however, collapsed in the 1990s with the mayorship of Diane Feinstein.

SYMBOLIC INCLUSION

The governing and regime coalition approaches offer two different means to interpret the ecological modernization efforts that were initiated on the local level, and then adopted by state governments, as well as by the federal government—as in the case of the Clean Air Act of 1990 (Gonzalez 2001a, chap. 6). Thus, public policies designed to force the development and implementation of air pollution control technologies could have been the result of public officials incorporating environmental and public health advocates in response to voter mobilization efforts around clean air issues. Alternatively, such advocates could have been incorporated, and ecological modernization policies could have been subsequently initiated, because segments of the public threatened to be politically and economically disruptive if their air pollution concerns were not addressed.

A third possible motivation underlying the incorporation of the leaders of public interest groups is to provide a democratic facade to the policymaking process. This incorporation would therefore be largely symbolic. The implication here is that the incorporation of public interest group leaders into the policymaking process has little or no substantive impact on policy outcomes. Instead, this incorporation occurs to communicate to the broader public that the public policy formulation process is reflective of various perspectives—thereby enhancing the legitimacy of said policies (Edelman 1977; Wynne 1982; Saward 1992).

Mark Dowie (1995), in his critique of large "mainstream" U.S. environmental groups, alleges that these groups have been knowingly incorporated on a symbolic basis. He specifically holds that the leaders of the major environmental groups prioritize organizational maintenance over achieving policy goals. Toward this end, environmental groups' leaders find it more important to be incorporated, or "close to power," than to "fight" for political goals, particularly since the former is a better fund-raising strategy.

The conclusions of Ronald Shaiko's (1999) study of five leading environmental groups is consistent with Dowie's argument that environmental groups' leaders generally prioritize organizational maintenance over achieving effective environmental protection. The focus of Shaiko's analysis is the relationship between the leadership of environmental interest groups and their members. He specifically seeks to understand the ability of the leaders of environmental interest groups and its membership to communicate on policy questions. This requires a two-part assessment. First, Shaiko analyzes the extent to which interest group leaders solicit their members for their opinion on various policy questions, and the extent to which institutional mechanisms exist within these organizations that allow members to communicate their policy preferences to the leadership. Second, he analyzes the ability and success of environmental interest groups to mobilize its membership on public policy questions. Shaiko utilizes data from five environmental interest groups to gain an understanding of the relationship between interest group leaders and their members: Sierra Club (SC), The Wilderness Society (TWS), National Wildlife Federation (NWF), Environmental Defense Fund (EDF) (now Environmental Defense [ED]), and Environmental Action (EA). The most significant conclusion drawn by Shaiko from his analysis is that environmental interest group leaders tend to prioritize organizational maintenance over political advocacy.[4]

According to Shaiko, a key reason for environmental interest groups' increasing emphasis on organizational maintenance is the public interest group milieu, which during the last thirty years has seen a substantial growth in the number of public interest organizations competing for members among a limited pool of individuals with the inclination and disposable income to pay membership fees. With an emphasis on organizational maintenance, environmental interest group leaders, to varying degrees, have come to view their members more as an economic constituency and less as a political constituency. This is best exemplified and reinforced by two trends among environmental groups: (1) the hiring of individuals outside of the environmental movement to be leaders of environmental interest groups for the specific purpose of organizational maintenance, and (2) the offering of perks to individuals to join or renew their memberships. Such perks include credit cards, posters, calendars, and magazine subscriptions. Moreover, Shaiko's analysis demonstrates that environmental interest group members tend to join these organizations largely for the tangible perks and less so for reasons related to political advocacy and public policy.

Next, I turn my attention to another interest group historically active on the issue of air pollution—business. Like that of environmentalists, the role of business in policymaking is contentiously debated. Unlike environmentalists, however, the question surrounding corporate America's political activity is not whether it has been influential over policy formation, but to what extent it influences state behavior. Is business's political power circumscribed by political officials drawing in broader public concerns and competing interest groups, or does the ownership and leadership of corporate America determine the state's agenda and how that agenda is addressed?

BUSINESS POLITICAL BEHAVIOR

PLURAL ELITISM

Up to this point in the discussion, I have written about business political power in general terms, and the relationship of the state, in broad terms, to economic elites. What specific form does business political behavior take, however, and what is the precise relationship between economic elites and government? Two ways to characterize business political behavior and the relationship of economic elites to the state are (1) plural elitism (Manley 1983; McFarland 1987; 1993; 2004; Lowery and Gray 2004) and (2) economic elite theory (Lamare 1993; 2000; Barrow; 1993, chap. 1; Domhoff 2002).[5]

According to the plural elite position, business political behavior is special interest politics. In other words, corporate firms' political activity is largely channeled through such narrowly construed organizations as business trade organizations, and their focus is largely their immediate political and economic self-interest (McConnell 1966; Grossman and Helpman 2001). Nevertheless, business groups are able to dominate those policy areas that correspond closely with their specific interests. This is because of the distribution of costs and benefits associated with participation in the policymaking process (Edelman 1964; Olson 1971), as well as because of business's command of the campaign finance system (Schlozman and Tierney 1986; Clawson et al. 1998; West and Burdett 1999), and businesspeople's control of society's productive forces (Lindblom 1977).

It is within this context of business firms and trade associations championing their narrow interests that public officials construe and enact broad policy proposals as indicated by the state autonomy/issue networks model. In this fragmented political milieu, Charles O. Jones (1975), for example, argues that political elites engaged in *speculative augmentation* to formulate and pass the Clean Air Act of 1970, whereby elected officials sought to outdo each other in the writing of this act to capture the increasing environmental vote. This political behavior was facilitated by the pro-environment social movement of the late 1960s and early 1970s (Tarrow 1994; Carter 2001). Echoing Robert Dahl's (1959) claim that "differences in the political behavior of businessmen may be almost as significant as similarities" (16), Bryner (1995), in his analysis of the formulation and passage of the Clean Air Act of 1990, makes specific reference to the putatively fragmented nature of business political activity and how this allowed the act to reflect multiple perspectives. He writes that:

Given industries' economic clout, ranging from honoraria paid to members to campaign contributions, one might expect that lobbyists could have freely worked their will in the legislative process. But business lobbying is rarely, if ever, united, since competitive pressures cut in many different directions. (135)

ECONOMIC ELITE THEORY

While plural elite theorists describe how individual corporate decision makers dominate specific and narrow policy areas, economic elite theorists contend that these corporate decision makers, along with other individuals of wealth, develop and impose broadly construed policies on the state. Additionally, while plural elite theory views the business community as socially and politically fragmented, proponents of the economic elite model hold that the owners and leadership of this community can be most aptly characterized as composing a coherent social and political unit or class.

As I noted in the preceding chapter, Clyde Barrow (1993) points out that "typically, members of the capitalist class [or the economic elite] are identified as those persons who manage [major] corporations and/or own those corporations." He adds that this group composes no more than 0.5 to 1.0 percent of the total U.S. population (17).[6] This group as a whole is the upper class and the upper echelon of the corporate or business community. The resource that members of the economic elite possess that allows them to exercise a high level of influence over government institutions is wealth. The wealth and income of the economic elite allow it to accumulate superior amounts of other valuable resources, such as social status, deference, prestige, organization, campaign finance, lobbying, political access, and legal and scientific expertise (Barrow 1993, 16).

Within the economic elite model, despite the segmentation of the economic elite along lines that are related to their material holdings, most policy differences that arise due to differences in economic interests can and are mediated. There are social and organizational mechanisms that exist that allow business leaders to resolve difficulties that develop within a particular segment and between different segments of the corporate community. For specific industries, or for disagreements between different industries, trade or business associations can serve as organizations to mediate corporate conflict. Social institutions, such as social and country clubs, can also serve as means through which to develop political consensus among the upper echelon of the business community on various economic, political, and social issues (Domhoff 1974). Michael Useem (1984), based on his extensive study of large American and British corporations, argues that corporate directors who hold membership on more than one board of directors tend to serve as a means through which the corporate community achieves consensus on various political issues (also see Mintz and Schwartz 1985).

On broad issues, such as air pollution, business leaders are also able to arrive at policy agreement and consensus through "policy-planning networks." According to G. William Domhoff, the policy-planning network is composed of four major components: policy discussion groups, foundations, think tanks, and university research institutes. This network's budget, in large part, is drawn directly from the corporate community. Furthermore, many of the directors and trustees of the organizations that comprise this policy-planning network are often drawn directly from the upper echelons of the corporate community and from the upper class. These trustees and directors, in turn, help set the general direction of the policy-planning organizations, as well as directly choose the individuals that manage the day-to-day operation of these organizations (Domhoff 2002, chap. 4).

Domhoff describes the political behavior of those members of the economic elite that manage and operate within the policy-planning network:

> The policy-formation process is the means by which the power elite formulates policy on larger issues. It is within the organizations of the policy-planning network that the various special interests join together to forge, however, slowly and gropingly, the general policies that will benefit them as a whole. It is within the policy process that the various sectors of the business community transcend their interest-group consciousness and develop an overall class consciousness. (Domhoff 1978a, 61)

Therefore, those members of the economic elite that operate within the policy-planning network take on a broad perspective, and act on behalf of the economic elite as a whole. Within this policy-planning network, members of the economic elite take general positions on such issues as foreign policy, economic policy, business regulation, environmental policy, and defense policy questions (Weinstein 1968; Eakins 1969; 1972; Kolko 1977; Domhoff 1978a, chap. 4; 2002, chap. 4; Barrow 1993, chap. 1; Gonzalez 2001a; 2001b).

This broad perspective also allows the policy-planning network to develop plans and positions to deal with other groups and classes. The network, for example, develops positions and plans concerning such policy areas as welfare and education. These plans can take several forms depending on the scope and level of the problems facing the business community and the state (Weinstein 1968; Eakins 1972; Domhoff 1978a; 1990; 1996; 2002; Barrow 1990; 1992; 1993, chap. 1; Dowie 2001; Cyphers 2002; Roelofs 2003).

Domhoff argues that the focal point in the policy-planning network is the policy discussion group. The other components of the policy-planning network—foundations, think tanks, and university research institutes—generally provide original research, policy specialists, and ideas to the policy discussion groups (Domhoff 1978a, 63). Policy discussion groups are largely composed of members from the corporate community and the upper class. Examples of policy discussion groups are the Council on Foreign Relations, the Committee for Economic Development, the National Association of Manufacturers, and the Chamber of Commerce. Overall, policy discussion groups are the arenas where members of the economic elite come together with policy specialists to formulate policy positions, and where members of the economic elite evaluate policy specialists for possible service in government (Eakins 1972; Domhoff 1978a, 61–87; 2002, chap. 4; Barrow 1993, chap. 1).[7] One example of a policy discussion group is the Clean Air Working Group. It was organized by the corporate community in the 1980s and focuses on the issue of federal clean air regulations (Gonzalez 2001a, chap. 6).

Certain environmental groups, in terms of their leadership and/or financing, have the characteristics of economic elite-led policy-planning organizations. These groups include the Sierra Club prior to the 1960s, the Save-the-Redwood League, and Environmental Defense. Environmental Defense, for instance, receives significant financing from large foundations, and it has several corporate executives on its board of directors (Dowie 1995, 58–59; 2001, 93; Roelofs 2003, 138–139). Susan R. Schrepfer, in her survey of the Sierra Club's early charter members, found that approximately one-third were academics, and "the rest of them were almost all businessmen and lawyers working in San Francisco's financial district" (1983, 10; also see Jones 1965 and Orsi 1985). The club was founded in 1892. Schrepfer goes on to explain that businesspeople continued to compose a substantial portion of the club's membership and leadership until the 1960s (1983, 171–173; also see Cohen 1988). Unlike the Sierra Club, the high level of economic elite participation on Save-the-Redwood League's governing council has been maintained throughout its history. The closed governance structure of the league created the "tendency for the council and board to be increasingly dominated by businessmen and patricians, while fewer academics were drawn into the organization's leadership in the 1950s and 1960s" (1983, 113). Through their financing and participation in such organizations businesspeople can gain knowledge and policy proposals on environmental issues. Economic elites can then use this information and these policy proposals in their efforts to shape public policies on environmental questions when deemed necessary (Gonzalez 2001; 2001b).

Economic elite-led policy discussion groups have also been formed for the purpose of shaping decision-making on the urban level. One prominent example of such an entity is the National Municipal League (Hays 1964; Domhoff 1978b, 160–169). From the nation-wide effort of this organization came the Progressive Era urban reforms of the civil service "to regulate personnel practices, competitive bidding to control procurement, the city manager form of government to systematize decision making, and at-large elections to dilute the voting power of the working classes" (Logan and Molotch 1987, 152).

REGULATORY POLICIES, ECONOMIC ACTIVITY, AND ECONOMIC ELITES

Gabriel Kolko (1977), in his work on Progressive Era politics, demonstrates that economic elites and large firms can and do benefit from government regulatory policies. Such policies can protect the value of investments, stabilize the operation of the market, and enhance the long-term profitability of capital. In the realm of urban politics, Marc Weiss (1987) shows how local zoning laws, and regulations regarding the building of housing and retail structures, were championed by large land developers beginning in the Progressive Era. They did so to protect land values and local investment climates.

Despite these seminal works, and others like them (e.g., Stigler 1971; Gordon 1994; and Higgens-Evenson 2003), Barrow (1998) points out that it is assumed by numerous scholars who study state behavior that a negative relationship exists within jurisdictions between business investment and regulatory rules applied to business activities (e.g., Poulantzas 1973; Offe 1984; Block 1987; Elkin 1987; Aronowitz and Bratsis 2002). Theorists that forward this view will normally link it to the "dependency principle," wherein it is posited that governments are reliant on private investment for a healthy economy and stable tax base (Offe 1974). Barrow explains that "in the literature on state theory, the operation of the dependency principle is always linked to a laissez-faire concept of the business climate and therefore to the basic presuppositions of neoclassical economic theory and the model of perfect competition" (111). Barrow goes on to point out that theorists of the state that argue the dependency principle generally equate a "favorable business climate" with

> "low" taxes (and therefore minimal state expenditures); low employee mandates such as minimum wages, unemployment insurance, workmen's compensation, and family leave; minimal social regulation and environmental protection; right-to-work legislation to protect a "free" labor market and correspondingly low wages. (111)

Hence, public officials, in order to attract investment to their specific nation, region, or locality, must provide investors the type of low tax and low regulation milieu called for in neoclassical economic theory (Bartik 1991; Fisher and Peters 1998).

The linkage of the dependency principle to neoclassical thought, however, is empirically unwarranted. Neoclassical assumptions about business investment prove to be poor predictors of where investment and economic growth in the United States occur. The General Manufacturing Climates, for example, was a ten year effort (1979–1988) to "operationalize neoclassical assumptions through an index that compares and ranks business climates in the 48 contiguous American states" (Barrow 1998, 112). It was sponsored by the Conference of State Manufacturers' Associations (COSMA). The General Manufacturing Climates used the "Grant Thornton index" to measure a favorable business climate based on "low wages, low union density, high work force availability (i.e., high unemployment), conservative state and local fiscal policies, and low state-mandated employment costs" (Barrow 1998, 112). In explaining the failure of this index to predict the national patterns of economic investment and growth, Barrow points out that its rankings

> were always inconsistent with perceived realities since North Dakota, Nebraska, and South Dakota ranked among the top three business climates year after year. Likewise, during the 1980s, California (30), Connecticut (35), and Rhode Island (37) ranked well below the median even though these states were in the midst of a robust expansion of business activity, employment, and personal income growth. (112)

Other studies testing the reliability of the Grant Thornton index found it significantly lacking in its ability to explain patterns of economic activity and investment in the United States (Lane et al. 1989; Kozlowski and Weekly 1990).

Barrow proffers a different way to conceptualize business climates and the role of governments in creating a "positive" investment climate. He specifically rejects the neoclassical approach, which conceptualizes economic activity as a series of "abstract market exchanges" that occur without effort, without friction, and without a legal framework. Instead, Barrow argues cogently that:

> In terms of constructing a concept of business climate, the taxes and fees that support public infrastructure, public education, and state-regulated employee mandates should all be regarded as transaction costs. They are part of the cost of creating and using markets. (117) (also see Ely 1914; Commons 1924; Coase 1960; Higgens-Evenson 2003)

Therefore, as opposed to viewing public infrastructures and various regulations as a drain on private firms and economic activity, such infrastructures and regulations should be viewed as a complex matrix that facilitates the operation of the market and economic activity. Moreover, this publicly provided matrix "subsidizes" business activity, since most state and local finances are generated through regressive taxes.

With this view of the relationship between regulatory frameworks and economic activity, the economic elite model would indicate that economic elites strive to implement those regulatory frameworks that protect and enhance economic activity and their market position. Historically, it has been demonstrated that economic elites will promote environmental regulations when they view it in their economic and/or political interest to do so. Casner (1999), for example, demonstrates that the Pennsylvania Railroad in the 1920s successfully advocated the regulation of water pollution from coal mines in Pennsylvania in order to protect its trains from said pollution.

Additionally, economic elites will utilize policy-planning networks to formulate the "appropriate" regulations, and to achieve political consensus on such issues. For example, to develop its expertise and to formulate specific policy proposals, the Clean Air Working Group, which began operating in the early 1980s, divided its "operations into roughly 10 separate teams handling the key clean-air issues" (Cohen 1995, 125). Richard Cohen (1995) goes on to report that the group's "weekly meetings, usually attended by more than 100 corporate lobbyists, often featured freewheeling discussions among different industry groups" (125–126). Therefore, when the need arose in the late 1980s to revise the Clean Air Act, the Clean Air Working Group had developed the knowledge and expertise to produce clean air policy proposals that were technically sound and agreed upon by its members. Moreover, given the needs of the business community in the late 1980s, this organization favored the strengthening of federal clean air laws through the Clean Air Act of 1990 (Gonzalez 2001a, 102).

STATE IMPERATIVES, ECONOMIC ELITES, AND THE CAPITALIST ECONOMY

Returning to the issue of the imperatives of the state as elaborated by Dryzek, the economic elite model would suggest that the state's imperatives are not determined within the state in response to different shifts in the operation of the political economy and/or public opinion. This view is implicit in the neo-Marxist view of politics (Barrow 1993, chap. 2; Dryzek 1996b; Aronowitz and Bratsis 2002), as well as in the state autonomy/issue networks model. Instead, it is economic elites, operating through policy-planning networks, that determine which issues within capitalism are to be addressed by the state and how.

Various works on U.S. environmental policies support this position. It was economic elites, for example, both within the lumber industry and outside of it, that developed a forest management approach during turn of the twentieth century that prioritized profitability, as well as forest maintenance, in order to cope with a long-running glut of the timber market. The development and dissemination of this approach, known as "practical forestry," was heavily subsidized by the federal government (Gonzalez 1998; 2001a, chap. 2). Moreover, William G. Robbins (1982) and Paul Hirt (1994) demonstrate that the management of the national forests has historically been dictated by the lumber industry. This industry makes its demands upon government in response to fluctuations in the timber market. In the case of the national parks, economic elites led in the creation of the National Park Service so tourism to the parks could be maximized. The national parks, in turn, became a more profitable outlet for capital investment for those economic sectors that economically benefit from this tourism (Runte 1997; Sellars 1997; Gonzalez 2001a, chap. 3; Barringer 2002). Moreover, it was live-stock firms, seeking land tenure stability, that led in the formulation of the federal government's policies managing grazing on the public grasslands (Foss 1960; Gonzalez 2001b).

CONCLUSION

The state autonomy/issue networks model offers one means through which to analyze state behavior in response to the various political controversies created by the operation of capitalism—including that of air pollution. According to this model of policy formulation, officials within the state take the political lead in addressing such controversies. Additionally, the incorporation of public interest advocates within the policymaking process can have a significant impact on the behavior of public officials. The incorporation of public interest advocates, however, can take different forms and have different affects on policy outputs. The incorporation of such advocates within the policymaking process could constitute a democratization of this process, represent co-optation, or symbolic inclusion.

The economic elite theory model offers a substantially different view of the policymaking process than that of the state autonomy/issue networks model. Here, economic elites, and not public officials, are at the center of the policymaking process. Operating through policy-planning networks, economic elites determine what types and which regulations will forward their economic and political interests. They, in turn, utilize their superior political means to ensure that the resources and organizational techniques that comprise the state are deployed to correspond to their preferred policies. In doing so, economic elites forward those public policies that they believe serve their interests and block those they perceive as not.

In the chapters that follow, I will describe how human-made air pollution has come into being and has changed over time, as well as how different actors have historically responded to this pollution. I begin in the next chapter with the late nineteenth century, when localized air pollution first became a significant problem for U.S. urban areas. It will be evident that the economic elite model offers more explanatory power than the state autonomy/issue networks model (or than pluralism) in analyzing the content and trajectory of government responses to air pollution.

THREE

The Politics of Air Pollution during the Late Nineteenth and Early Twentieth Centuries

The Failure of Technology

SAMUEL HAYS (1987; 2000) argues that the current U.S. air pollution abatement regulatory regime, along with other environmental regulatory regimes, are the result of the expansion and rising affluence of the U.S. middle class during the post–World War II period. With its increasing economic security, substantial segments of this middle class, according to Hays, moved away from their almost exclusive concern with economic issues, and came to prioritize politically the environment for its aesthetic and salutary qualities (also see Inglehart 1977).

David Stradling (1999), however, demonstrates that the clean air movement can be traced back to the late nineteenth century. He, nevertheless, also argues that the political impetus underlying this movement was primarily a middle class concerned with the adverse effect that air pollution had on the aesthetics of the urban milieu and on the health of urban residents. Given this class make-up of the clean air movement, Stradling argues that its primary focus and result is the effort to control air pollution through the development and application of technology. In contrast to the position taken by Hays and Stradling, I hold that the primary political energy underlying the historic development of the U.S. clean air regulatory regime is provided by economic interests that monetarily benefit from local economic and population growth, and, subsequently, increasing land values as well as an expanding consumer base.

COAL AND AIR POLLUTION

Severe air pollution afflicted many major U.S. cities during the late nineteenth and early twentieth centuries. This was the result of burning coal. Coal was the fuel of the second industrial revolution. Alfred Chandler (1972) argues that coal is what made this industrial revolution possible. He explains that the speed at which heat could be generated through the burning of coal facilitated the productivity and efficiency that was achieved through economies of scale. Moreover, coal fueled the North American railroad network that allowed firms to access the resources and markets necessary for the profitable operation of economies of scale.

Coal, however, when combusted generates smoke. This smoke largely takes the form of sulfur dioxide. Soft, or bituminous, coal is an especially polluting form of coal when burned. Thus, while coal, and especially soft coal, was at the center of the U.S. economy, so was smoke, i.e., air pollution. There were certain factors however that exacerbated the concentration of smoke during this period. First, the concentration of transportation nodes in particular cities, and, second, the concentration of industrial facilities in these same cities. Stradling (1999) succinctly describes the environmental, existential, health, and economic problems facing U.S. industrial cities in the late nineteenth and early twentieth centuries resulting from the acute air pollution generated by the concentrated burning of coal, particularly the soft variety:

> As smoke hung over the city in dense clouds and fell in the form of black, acidic soot, it posed a multitude of serious problems for industrial cities. The dark clouds obstructed vistas and blocked sunlight, adding gloom to the city and often necessitating the use of artificial light even in daylight hours. The soot, much more troublesome than the smoke itself, helped make industrial cities filthy, inside and out, as it clung to clothes, furniture, drapes, and rugs. Soot darkened buildings and its acids ate into stone and steel alike. Merchants complained of unsold goods spoiling on store shelves, housewives protested against never-ending cleaning, and doctors warned that urban lungs were harmed by smoky air. (3)

Therefore, cities, as aggregations of productive capital and labor, are centers where natural resources are converted into commodities and services (Harvey 1985). Coal, as the natural resource that powered these centers and the means of transportation that linked them to each other and the hinterland, created environmental conditions that severely degraded the life of urban residents (Vietor 1980).

What was the political response to the pall of smoke that engulfed numerous U.S. industrial centers beginning in the late nineteenth and through to the middle of the twentieth century? Matthew Crenson (1971), in his classic study of air pollution in Gary, Indiana, describes how this city responded to the problem of air pollution during this period (also see Greer [1974] and Hurley [1995]). The city of Gary dealt with its severe air pollution problem, created mostly by its U.S. Steel plant, by politically ignoring it. Thus, Gary, Indiana, kept the smoke problem associated with local industrial facilitates and railroad lines off of the political agenda. Many cities did pass regulatory legislation to control smoke emissions, but these were largely unenforced. Even large cities, such as Chicago, Pittsburgh, and St. Louis, which were large users of soft coal, failed to effectively address the smoke that filled their air (Stradling 1999; Dewey 2000).

Why? The failure to deal with smoke during the late nineteenth century and throughout the early twentieth can be attributed to primarily three factors: (1) the goals of local growth coalitions; (2) the unwillingness of railroad firms to ecologically modernize their lines; and (3) the inability to control coal-generated smoke through the deployment of technology. I examine the air pollution politics of Chicago during the period in question to bring these factors into relief.

LOCAL GROWTH COALITIONS AND SMOKE

Chicago was one of a number of American cities during the late nineteenth and early twentieth centuries that suffered from the severe inundation of smoke from the burning of soft coal. Cities like Chicago, Pittsburgh, St. Louis, Birmingham, and Cincinnati were historically reliant on bituminous or soft coal as a source of cheap fuel. Until the 1930s, cities in the Northeast, such as New York, Boston, and Philadelphia, tended to produce substantially less smoke than those in other regions east of the Mississippi River, because in large part they had relatively inexpensive access to cleaner burning anthracite or hard coal. Nevertheless, by the 1930s these cities, too, (New York City in particular) had significant air pollution difficulties from the burning of soft coal (Williams 1997; Dewey 2000, 113–114; Melosi 2001).

In urban centers, excessive smoke was viewed as an economic negative by many local business elites, proponents of local growth, and local business associations. As one historian who has studied U.S. air pollution politics explains, "Local business leaders and civic boosters who sought smoke controls, . . . viewed smoke as a threat to growth rather than to health." He elaborates that "such smoke fighters offered statistics showing that St. Louis had suffered six million dollars in smoke damage in 1906, or that smoke had cost Cleveland six million, Cincinnati eight million, and Chicago fifty million dollars in 1911 alone" (Dewey 2000, 25). In Pittsburgh, the Chamber of Commerce's Committee on Smoke Prevention declared in a 1899 report that the "smoke has become such a nuisance in our City as to call for some very decided action, if we hope to claim for Pittsburgh the advantage of its being a good place for business" (Chamber of Commerce of Pittsburgh 1900, 60). Therefore, "because of such worries, local business interests and associations were active in pushing for smoke control in many cities" (Dewey 2000, 25), such as New York, St. Louis, Pittsburgh, and Chicago (Grinder 1978; Rosen 1995; Stradling 1999; Dewey 2000; Gugliotta 2003).

Within this context, the case of Chicago takes on theoretical and historical significance, because members of its local growth coalition undertook direct action to effectively control the emission of smoke. It is by understanding why these actions failed to abate smoke from coal that we can see why other cities failed during this time to translate concerns about smoke into effective smoke abatement.

RAILROADS AND AIR POLLUTION

Railroads, which used largely soft coal, were a source of smoke for most major cities. It was a particularly significant source, however, for Chicago. This is because the city was a primary North American railroad hub. As historian William Cronon (1991) outlines, Chicago was the key entry point for Northeastern and European capital, goods, and services into the Midwest and West. In turn, the city was the central means through which raw materials from the Midwest and West found their way into Northeastern and European markets. Moreover, Chicago itself became a major center of industrial production, and the railroads serviced this substantial manufacturing base (Mayer and Wade 1969; Keating 1988).

Given the significant railroad traffic that Chicago experienced, it was a logical place where a political effort would emerge to control the smoke generated by this traffic. This is for two reasons. First, and most obviously, with such a high number of railroad locomotives moving into and out of the area, these locomotives would produce a large amount of smoke. Second, with so many railroad firms invested in the Chicago area, if the city did impose expensive smoke abatement measures, it might be too costly for these firms to move out of the area despite the costs associated with these measures. As one president of the Chicago, Burlington, and Quincy railroad put it, "Railroads are fixtures; they cannot be taken up and carried away" (as quoted in Cronon 1991, 83).

Therefore, shortly after the railroads in New York City were electrified, and the Anti-Smoke League, a woman's organization, began a campaign protesting the smoke emitted by the Illinois Central lines running through the upscale lake front and Grant park areas (Stradling and Tarr 1999, 693), the Chicago Association of Commerce (CAC) conducted a study of the smoke generated by the railroads in the area. The CAC was a local organization, composed, as Stradling (1999, 126) explains, of "many of the city's most important businessmen." Given its composition and work on numerous policy issues (Walker 1941),[1] the CAC can be categorized as a policy discussion group for the Chicago business community. In its initial report, completed in 1910, the Association concluded that "electrification" of the railroad lines in the Chicago area "should be no great difficulty" and that it could be successfully accomplished "within a reasonable time." The CAC report went on to assert that electrification of the city's railroads "is not only practicable, but will be of great advantage to Chicago, and it is recommended for [immediate] execution" (Chicago Association of Commerce 1915, 20).

The Chicago Association of Commerce, however, did not initially publish its findings. Instead, it presented them to numerous railroad executives to seek their support (Stradling and Tarr 1999, 694–696). This noncon-frontational tack taken by the CAC is to be expected. Because while Chicago's local business community had an interest in controlling smoke, its fortunes were simply too dependent on the railroad firms for its members to take a politically antagonistic attitude toward them.

Allan Pred (1966), in his comprehensive analysis of the rise of the U.S. city-system during the nineteenth century, makes special note of the importance of the railroad to this system and its expansion, and how Chicago assumed an important place within the U.S. city-system during this period precisely because numerous railroad lines were placed there. Pred explains that "the paramount importance of initial rail and terminal facility advantages to urban-industrial growth is best exemplified by Chicago." He then goes on to describe in detail how the growth of Chicago during the latter half of the nineteenth century can be directly attributed to the railroad:

> In 1860 Chicago was a city of moderate size, with less than 5,400 workers employed in printing, carpentering and cabinetmaking, boot and shoe production, baking, coopering, tailoring, brickmaking, and other small-scale industries that catered primarily to local markets. Even the city's 16 foundries and machine shops, with almost 600 laborers, were occupied largely with fulfilling special local orders; and not one manufacturing establishment in the young metropolis had a working force in excess of 200. Within fifty years industrial employment in the physically expanded metropolis had grown to more than 325,000, many of them in the categories whose firms frequently served distant markets (e.g., the machinery and foundry product industry had 41,492 employees; the slaughtering and meat packing industry had 27,083; and the iron and steel industry, 16,730 employees). Chicago's phenomenal rise was foreshown by 1860, when the city had emerged as the nation's most important railroad center, a terminus for 11 trunk roads and 20 branch and feeder lines. (54)

Between 1870 and 1890, "the mileage of railroads entering Chicago had increased 370 percent; their tonnage 490 percent" (Hoyt 1933, 142). Additionally, as one student of Chicago's economic growth notes, "There was a rush of manufacturing concerns to locate in the Chicago area to obtain the advantage of its superior terminal facilities and favorable railroad rates" (Hoyt 1933, 143). Thus, between 1884 and 1890, the value of manufactured products increased during this period from $292 million to $664 million (Hoyt 1933, 144).

The significant railroad investment placed in Chicago and its subsequent economic growth are not only attributable to its location adjacent to Lake Michigan and position in relation to New York City. In addition to these factors, Chicago was a good place for capital investment during the middle and late nineteenth century because local actors created the physical and political milieu conducive for such investment.

With the clearing out of the Native American population, and cheap credit, Chicago during the 1830s became part of the broader effort to profit from the expected colonization of the Old Northwest territory. Cronon (1991) explains that

> the mid-1830s saw the most intense land speculation in American history, with Chicago at the center of the vortex. Believing Chicago was about to become the terminus of a major canal, land agents and speculators flooded into town, buying and selling not only the empty lots along its ill-marked streets but also the surrounding grasslands which the Indians had recently abandoned. (29)

Therefore, once the Indian population was moved out, thereby opening the Old Northwest territory to be integrated into the U.S. and European economies, inhabitants and investors in significant numbers came to the Chicago area to economically benefit from an anticipated land boom. The views of this class of Chicago residents and investors came to dominate the city's politics (Belcher 1947; Keating 1988).

This pro-growth political outlook led to various projects designed to attract capital investment to the area. It, for example, facilitated the building of a canal in the 1840s that connected Lake Michigan to the Mississippi River watershed, thereby making the Chicago area an important North American transport point (Belcher 1947, 34–35; Cronon 1991, 32–33). Moreover, in the late 1840s, it was leading members of Chicago's real estate interests that built the first railroad line running to Chicago (Belcher 1947, 125–131; Cronon 1991, 65–67). In the 1850s, Illinois Senator Stephen Douglas, who himself held substantial tracts of land in the area (Belcher 1947, 125–126), successfully lobbied to have the first federal railroad land grant run to Chicago (Cronon 1991, 68–70).

This congenial attitude toward investment and economic growth in Chicago paid off well for those that owned substantial amounts of land in the area. Between the years 1833 and 1910, land value within the incorporated area of Chicago had grown from $168,000 to $1.5 billion. By 1926, the eve of the Great Depression, total land value in Chicago equaled $5 billion (Hoyt 1933, 470; also see Keating 1988).

Therefore, the CAC's first report on railroad-generated smoke, and the recommendation to electrify the railroad lines running through the area, were not a break from the congenial political attitude that Chicago's growth coalition had historically maintained toward the railroads, specifically, or capital, generally. Instead, the CAC report should be viewed as an attempt to initiate a process, whereby it was hoped that the major railroads would eventually agree to electrify its Chicago lines.

Reflective of this approach to the railroad-related smoke problem, even before the CAC report was completed, the CAC executive committee began to meet privately with railroad officials to discuss the idea of electrification. Historians Stradling and Joel A. Tarr (1999) report on the response in these private meetings of one of those railroad firms, the Pennsylvania, to the CAC project of railroad electrification. The executives of this firm strongly rejected the concept of electrification in Chicago. A General Manager of the railroad, G. L. Peck, stated to the CAC executive committee that railroad electrification in Chicago "was out of the question." Another Pennsylvania Railroad executive at the same meeting pointed to what he described as the "prohibitive cost of electrification and the impracticability of substituting coke or anthracite coal" for soft coal. Peck also warned the CAC executive committee that "the operating officers of the railroads" would oppose its report. In reporting the results of the meeting to a Pennsylvania vice president, Peck explained that the CAC "Executive Committee decided to suppress the report" (as quoted by Stradling and Tarr 1999, 695). Stradling and Tarr (1999) note that "after meeting with other railroad officials, the CAC agreed to shelve its study" (695).

The CAC, however, not only agreed to end its efforts at having Chicago trains run on electricity but it aided the railroad industry in putting the issue to rest politically. Under the auspices of the CAC, a committee was formed to again study the smoke from railroads in the Chicago area. This committee was composed of various officials from the railroad industry. Executives from such railroads as the Chicago & North Western, the New York Central Lines, and the Chicago, Burlington & Quincy served on the committee. Additionally, individuals economically tied to growth in the Chicago area were also members of "The Chicago Association of Commerce Committee of Investigation on Smoke Abatement and Electrification of Railway Terminals." These individuals were bank executives, the senior partners of law firms, and one senior partner of an architect firm. Finally, an executive from the coal industry was also a member of the committee (Chicago Association of Commerce 1915, v).

The new report, released in 1915, downplayed the contribution that railroad locomotives made to Chicago's air pollution. Specifically, the report

attributed approximately 22 percent of the area's air pollution to steam locomotives (Chicago Association of Commerce 1915, 173). This figure substantially differed from that arrived at earlier by the city's smoke inspector and the
CAC's first report. The city's inspector estimated that steam locomotives
accounted for 43 percent of the smoke generated in Chicago (Stradling 1999,
128), whereas the Association's initial report estimated that they were
responsible for anywhere from 30 to 50 percent of smoke emitted in the area
(Chicago Association of Commerce 1915, 22–23).

Moreover, the thrust of the report served to present the railroad firms'
point of view on the air pollution question as it related to steam locomotives,
and this is in part a reflection of the fact that the committee relied very heavily on the railroads for its data (Stradling 1999, 129). First, the 1,052-page
report focused almost exclusively on the costs associated with the electrification of trains within Chicago. Second, it almost totally ignored the costs associated with air pollution and hence the benefits potentially accrued with the
abatement of such pollution.[2] Third, the report began with a history describing how the growth of the city can be directly related to railroad investment
in the area (Chicago Association of Commerce 1915, 11–16). The authors of
the report explained that "wherever a railroad has chosen to marshal its cars,
there the city has ultimately crowded in." They went on to write that "the
railroads have been an impelling force which has aided in the city's legitimate
development" (Chicago Association of Commerce 1915, 15).

With the committee's emphasis on the costs of electrification and its uncertainty regarding the benefits associated with air pollution abatement, along with
its view that steam locomotives accounted for only approximately one-fifth of
the smoke in the area, unsurprisingly the committee advised against "immediate
or general electrification of railroads for the purpose of eliminating their part in
air pollution." Additionally, it advised against any action abandoning "the use
of Illinois and Indiana coal" (Chicago Association of Commerce 1915, 1052),
which is generally of the bituminous or soft coal type (Chicago Association of
Commerce 1915, 16–17; Platt 1995, 74). The report's release in 1915 effectively
ended the possibility of the government mandated electrification of the railroads
within the Chicago area (Stradling 1999, 130).

It is not that the owners and managers of industry and the railroads were
indifferent to the difficulties associated with air pollution. For example, historian Scott Dewey (2000, 25) explains that "in Pittsburgh, Andrew
Carnegie, Henry Clay Frick, and George Westinghouse all shared with the
local chamber of commerce an interest in smoke control" (also see Grinder
1978). Moreover, in 1927 the Illinois Central did voluntarily electrify its suburban lines in Chicago (Stradling 1999, 130). Nevertheless, the costs and
uncertainties associated with pollution abatement technologies led industry
and the railroads throughout the late nineteenth and early twentieth centuries to oppose government enforced pollution abatement regulations.

The issue of cost was particularly paramount for the railroad industry. In the case of New York City, railroad firms electrified their lines because the smoke emitted from coal burning locomotives proved to be a substantial safety hazard within the tunnels in and around the city—with at least one major railroad accident within a tunnel attributed to locomotive-generated smoke. Also, the Pennsylvania Railroad ran many of its trains in Philadelphia on electricity, beginning in the 1910s, because space in its crowded Philadelphia facilities could be more efficiently used with electric locomotives (Stradling 1999, 112–114). While electric locomotives proved to be clean and efficient, the railroad industry as a whole was heavily vested in steam locomotives. One expert estimated that in the early part of the twentieth century the railroads owned approximately 70,000 steam locomotives, representing a total investment of $1.4 billion (Crawford 1913). Thus, in electrifying their lines nationally, railroads would not only have to absorb the considerable costs of doing so but these firms would have to effectively abandon their significant investment in steam locomotives. In light of these factors, it is not surprising that the railroads consistently opposed any attempts to forcibly electrify their industry (Stradling 1999, chap. 6). Given this strong opposition, those economic elites in Chicago interested in cleaning the city's air through the mandated electrification of those locomotives running through the area backed down. This backpedaling is reflected in the second report put out by the Chicago Association of Commerce.

THE LIMITS OF TECHNOLOGY IN AIR POLLUTION CONTROL

With regard to air pollution from factories, not only was there considerable costs associated with the installation, maintenance, and usage of pollution abatement technologies but the existing technological approaches to air pollution abatement were in many cases unreliable. Moreover, the costs and uncertainties associated with air pollution abatement technology gave rise to free rider problems. The difficulties of controlling coal-generated smoke through the usage of technology becomes most apparent in light of an effort to control such smoke in anticipation of the Chicago World's Fair in 1893.

Leading members of Chicago's business community in January 1892 formed the Society for the Prevention of Smoke. As historian Christine Rosen (1995, 358) points out, "The founders of the Society were prominent local businessmen, . . . and all but one of whom was also a Director" of the fair, otherwise known as the Colombian Exposition. Founders of the Society included Bryan L. Lathrop, the organization's president, and a real estate developer and investment banker. Also on the Society's board of directors were Samuel W. Allerton, a banker and cattle rancher, who helped found Chicago's Union Stockyard Company and was a director of a local street railway company, and James W. Scott, publisher of the *Chicago Herald*. When

Scott retired from the Society, he was replaced by Owen F. Aldis, a lawyer and real estate developer (*Biographical Dictionary* 1892, 55–56, 281–282; Rosen 1995, 358). As the General Director of the Colombian Exposition explained, the objective of the Society was to "get rid of the smoke nuisance before the world's fair opens" (as quoted in Rosen 1995, 359).

Initially, it sought to eliminate smoke from the central city area through voluntary cooperation. Toward this end, the Society hired five engineers. With these engineers, the Society would advertise those air pollution abatement devices that were believed to be effective. Additionally, the Society's engineers would inspect buildings in the central city area, and make recommendations through detailed reports to the owners of buildings that were guilty of excessive air pollution. These reports would advise owners what technologies and changes to design would be required to reduce the emission of smoke (Rosen 1995, 359–360).

Utilizing this approach of voluntary cooperation, the Society achieved moderate success. By July 1892 it found that approximately 40 percent of offending buildings had followed its engineers' advice and had substantially reduced smoke. Of the remaining 60 percent, the Society concluded that 20 percent were not using their abatement equipment correctly or were not properly stoking their fires, while 40 percent did not comply at all (Rosen 1995, 360).

To address those businesses that were not complying with the Society's efforts, beginning in August 1892, it sought to force compliance through the courts. Using an anti-smoke ordinance enacted in 1881 but heretofore unenforced, the Society began to cite those it deemed in violation of the law. It even hired its own attorney to prosecute the cases, and it was given the power to select the judges before which it would try its cases (Rosen 1995, 376). As Rosen (1995, 376) explains, contemporary critics did complain that this activity "represented a legally questionable appropriation of police power by a private organization." Nonetheless, throughout its period of operation the Society initiated 325 suits against alleged offenders (Rosen 1995, 377). With a deepening depression in the 1890s, some high profile defeats handed to it by juries, and the opening of the World's Fair, the Society ended its efforts at smoke abatement in August 1893 (Rosen 1995, 380–382).

A significant hurdle for those who sought to reduce smoke from steam generating boilers was the limited nature of the technology available to abate pollution from these sources. As explained by Rosen (1995, 359), "In addition to its cost, smoke abatement equipment was often difficult to install." Moreover, "Devices that worked well in one building often failed completely in another." It is entirely possible that many of those 20 percent that the Society found to be in partial compliance with its efforts had installed pollution abatement equipment, but such equipment did not work effectively in the given circumstance.

Perhaps the most significant obstacle to reducing smoke from coal-burning boilers was that the reduction of smoke required a significant amount of labor and expertise. Stradling (1999) points out that the "proper design and installation of equipment in itself did not ensure smokelessness, however, as great difficulty came from incompetent operation of furnaces, both new and old." He goes on to explain that experts of the time held that "the key to smokelessness . . . was proper handling by the firemen and the engineers" of boilers (80). The firemen and engineers responsible for stoking boiler fires, for example, had to break up coal into smaller pieces in order to reduce the output of smoke. Additionally, they had to burn lighter fires. Both of these factors would make the work of firemen and boiler engineers more difficult and labor intensive (Rosen 1995, 372). This situation was exacerbated by the fact that the fireman position was relatively low paid and viewed as an unskilled position. Stradling (1999) points out that "many businesses did not even attempt to attract educated and skilled men to work and remain in boiler rooms, for most plant managers continued to assume the positions required only strong backs and continued to pay their firemen as if they were expendable" (80).

With the expense and uncertainty associated with pollution abatement technologies, and the significant costs associated with the effective usage of such technologies, free riding was an obvious problem with the mandating of technology to affect smoke reduction. This problem could only be overcome through a command-and-control strategy undertaken by government, or a quasi-government agency—as Chicago's Society for the Prevention of Smoke became when it was granted enforcement and prosecutorial powers.

This, however, would create two apparent difficulties. First, such a command-and-control approach would require substantial expenditures, since not only would such an effort necessitate inspections to ensure that sources of air pollution installed the requisite equipment but enforcing agencies would also have to monitor these sources to ensure that such equipment would be properly and assiduously used. Second, if command-and-control efforts would be carried out on a city by city basis, this would serve to deter capital investment from those cities undertaking these efforts.

These factors contributed to the failure of the clean air movement to produce cleaner air during the late nineteenth and early twentieth centuries. This failure can at least in part be attributed to the fact that the economic elites that were pushing for air pollution abatement were seeking to do so exclusively through technology. Ultimately, it was not until the cleaner fuels of oil and natural gas became economically viable that several metropolitan areas that were also home to industrial centers had cleaner air (Sanders 1981; Tarr 1996, chap. 8; Stradling 1999; Dewey 2000; Laird 2001, 114).

THE BROADER PUBLIC?

Outside of the business community, what role did the broader public play in the clean air movement during the late nineteenth century and into the middle of the twentieth? In terms of an organized response, a relatively small one. In other words, to the extent that the urban working class was unhappy with the quality of air enveloping many urban environments, this unhappiness did not take any specific organized form. Or more simply put, throughout this period of time no mass-based organization came into being that mobilized the public on the issue of air pollution.

In light of the significant air pollution problem that plagued numerous U.S. cities from the late nineteenth century and into the middle of the twentieth, why did the working class not mobilize against this obvious aesthetic blight and health hazard?[3] The group mobilization incentive structure outlined by Mancur Olson (1971) offers part of the explanation as to why, even in the face of persistently poor air quality, the mass of urban residents remained quiescent on the issue of air pollution. The symbols emanated with public officials' rhetoric, the enactment of regulatory legislation, and the creation of underfunded smoke inspection agencies also contributed to the public's relative political passivity on the issue of air pollution.[4] These symbols would communicate to the public that something was already being done to address the issue of air pollution and that they need not spend their time and energy attempting to overcome the collective action barriers inherent in the mobilization of large groups (Edelman 1964; 1988; Cahn 1995).[5]

In writing about air pollution politics during the post–World War II period leading up to 1970, Dewey (2000) identified a particular trend. He explains that there was a "traditional air control cycle—a flurry of public interest stifled by bureaucratic inertia" (116). Thus, as indicated by Dewey, public concerns over air pollution in such cities as New York would be periodically aroused by some dramatic air pollution event, such as a summer of very poor air quality. These concerns would be aired through letters to the editors, newspaper articles, and the occasional public demonstration. In response to this public grousing, public officials would promise action, perhaps pass legislation, and often undertake well-publicized but essentially token enforcement efforts.[6] Despite this, normally little of substance was accomplished in terms of effective air pollution control. This was a common pattern of political activity as it related to the issue of air quality throughout the late nineteenth and early twentieth centuries (Stradling 1999; Dewey 2000).

Upper Class Women and Air Pollution

Apart from local chambers of commerce and particular economic elite-led organizations, organized opposition to air pollution during the late nineteenth and early twentieth centuries came from women's organizations.[7] Examples of women's groups that took active roles during this time on the air pollution issue in their respective cities were the Women's Club of Cincinnati, the Wednesday Club of St. Louis, and the Ladies' Health Protective Association of Pittsburgh. These groups' most significant contribution to the clean air movement was to draw public attention to the air pollution problem (Grinder 1978; Platt 1995; Flanagan 1996; Stradling 1999; Dewey 2000).

These women's groups were not mass organizations, with membership largely limited to the hundreds and very low thousands. Moreover, such groups had few working class members. Dewey (2000) explains:

> Working class wives and mothers might have favored smoke abatement . . . , and they almost certainly hated the smoke that ruined their laundry, befouled their homes, threatened their families' health, but most of them probably did not have the time or energy left over from their daily tasks to attend meetings or organize clubs. (25)

Instead, the women's organizations that took up the smoke question were in large part made up of upper-class women (Grinder 1980; Stradling 1999; Dewey 2000). For example, in writing about the Ladies' Health Protective Association, Angela Gugliotta (2000, 173) takes note of the class background of this group when she writes that in 1892 "the elite women of the Ladies' Health Protective Association took the decisive public action against smoke" in Pittsburgh. Dale Grinder (1980), in his work on the politics of smoke between the Civil War and World War I, acknowledges that the leadership of anti-smoke women's organizations was largely drawn from the upper strata of society. Grinder (1980) writes that "since only those with leisure time could devote themselves to most reformist causes, upper-middle-class women directed the clubs' anti-smoke efforts." He adds:

> Leadership in the local smoke abatement campaigns included Mrs. John B. Sherwood of Chicago, Mrs. Charles P. Taft of Cincinnati, Miss Kate McKnight of Pittsburgh, and Mrs. Ernest R. Kroeger of St. Louis, all members of the upper middle class, if not the social register [a publication identifying members of the U.S. upper class].

Moreover, "Both the Wednesday Club of St. Louis and the Twentieth Century Club of Pittsburgh, in the years prior to World War I, constructed clubhouses that indicated a membership of great wealth" (88).

G. William Domhoff, who has written extensively on the political activity of the U.S. upper class and corporate elite (1967; 1970; 2002), explains that members of the feminine portion of the upper class have historically volunteered in significant numbers in efforts that often deal with social ills (2002, 54–56). Drawing upon sociological work done on upper class women, Domhoff (2002) points out that these women view their civic and charitable work "as a protection of the American way of life" (55). In the case of air pollution, upper-class women were seeking to aesthetically improve the urban milieu through the application of technology, but their

critiques of urban air quality did not explicitly or implicitly critique industrial capitalism nor the socioeconomic structure that it yielded. In the following, Stradling (1999) describes the political outlook of anti-smoke
activists at the turn of the century:

> Progressive reformers, including anti-smoke activists, rarely offered a com-
> prehensive critique of the industrial order that lay at the root of the diverse
> problems they hoped to solve. Some reformers did organize against specific
> industries and even specific companies, but for most progressives the object
> of reform was to preserve the industrial system that had so enriched their
> communities and themselves. (2)

Given the upper class position of the women leading the anti-smoke movement, such an outlook should not be surprising.

CONCLUSION

The central argument of this chapter (and book) is that locally oriented economic elites have historically provided the key political capital forwarding
the ecological modernization of U.S. society as it relates to the issue of air
pollution. This position is predicated on two factors. First, that air pollution
is perceived as an economic negative by these elites. Indeed, as noted above,
members of local growth coalitions have historically emphasized the negative
economic impact of such pollution. The second factor my argumentation is
based upon is that locally oriented economic elites have viewed technological controls as an appropriate response to the economic negative of air pollution. Such a response can help manage air pollution without directly affecting the amount of economic growth that accrues in an urban region.

As I explained earlier, the case of Chicago during the late nineteenth
and early twentieth centuries takes on specific historical and theoretical significance. It was here where locally oriented economic elites sought to trans-
form their concerns about air pollution into regulatory policies. These policies sought to address the acute air pollution of the city through the
deployment of technology. The inherent political and technical difficulties,
however, associated with such technology led these elites to end their efforts
to control smoke. The political difficulties resulted from the opposition of
railroad firms to the forced electrification of their lines. The available technology to control smoke created by the burning of coal was of limited use
and expensive to employ. Given these political and technical difficulties,
areas like Chicago, which suffered from extremely poor air quality during the
late nineteenth and early twentieth centuries, simply lived with the severe
air pollution associated with economic activity and growth. Cities such as
New York, Pittsburgh, and St. Louis were just as pro-growth as Chicago during this period, and continue to be so (Belcher 1947; Pred 1966; Eisinger

1988; Keating 1988; Fainstein 2001). Instead of taking effective action against air pollution, and negatively affecting local investment levels, these cities, as a means of pacifying the public on the issue of air quality, simply passed smoke ordinances that went largely unenforced. Other symbolic actions undertaken by political elites to assuage the public on the question of air pollution included token enforcement efforts and rhetorical commitments to smoke abatement.

Of course, in relying upon technology to control smoke from coal-burning local growth elites eschewed more direct and effective methods to abate air pollution. A more wide-ranging and effective approach to air pollution abatement would have been to seek to balance the economies of scale achieved through the concentration of production and transportation facilities versus the clean air needs of particular communities. By spreading production and transportation facilities in this way, production efficiency would only be minimally affected, but cleaner air for urban areas could be achieved. Additionally, such an approach to the management of national investment patterns would have sought to place the nation's production and transportation infrastructure close to sources of clean fuel, such as anthracite or hard coal, natural gas, and oil, and away from dirty ones, such as soft coal. Significantly, job creation on a national scale would not have been adversely affected by such a distribution of production and transportation facilities (Molotch 1976; Simonis 1989; Crowley 1999).

Therefore, in leading the clean air movement during the late nineteenth and early twentieth centuries, both male and female economic elites, shaped the debate around air pollution to serve the broad interests and preferences of the U.S. economic elite. In other words, this movement was largely limited to advocating for the reform of industrial capitalism through the application of technology, and not by challenging the prerogatives of capitalists to invest when and where they wanted, nor questioning the practice of opening localities to unlimited levels of investment. Thus, the clean air movement of this period promoted a relatively limited type of reform (Christoff 1996; Dryzek 1997, chap. 8; Neumayer 2003), and one that failed. As noted earlier, it was not until oil and natural gas became more widely available as a source of industrial and transportation fuel in the post–World War II period that numerous cities experienced significantly cleaner air.

The ecological modernization efforts during the late nineteenth and early twentieth centuries were not undertaken to develop a more harmonious relationship between capitalism and the environment, or even between capitalism and human health, as indicated by much of the literature on ecological modernization theory (Mol and Sonnenfeld 2000; Young 2000; Mol 2001). Instead, such efforts were primarily undertaken to make industrial production and railroad transportation more congenial to the economic interests of one segment of the U.S. capitalist class—local growth coalitions. It is when we take this into account that we can understand why ecological modernization efforts left many U.S. cities during this period with persistently poor air quality (Rosenbaum 1998; Andrews 1999).

In the following chapters, I analyze the rise of the automobile, its contribution to air pollution, and why technology came to be the answer to this pollution. Despite the different source of pollution and the different time frame—the 1940s to the present—the politics surrounding automotive air pollution are very similar to those described in this chapter. Instead of Chicago, however, the historically significant events leading both to the dominance of the automobile as a local means of transportation and the effort to manage automotive pollution through technology occurred in Los Angeles.

FOUR

Real Estate and the
Rise of the Automobile

IN THE LAST CHAPTER, we observed how Chicago became a central node of the U.S. economy. During the middle and late nineteenth century it became the key transport point for the middle West, as well as a major area of manufacturing. Given that soft coal powered both of these activities, Chicago, and other metropolitan areas reliant on soft coal, had a significant air pollution problem. By the middle of the twentieth century, however, air pollution from the burning of coal began to substantially abate. Beginning in this period, a new source of air pollution started to emerge. This air pollution did not derive from a central energy source powering various facets of the economy, as in the case of coal. Instead, it was rooted in a transportation system that came to be the dominant means of rapid transit in U.S. cities. At the center of this system is the automobile.

A key political force historically underlying the broad adoption of the automobile in the United States has been the core groups within local growth coalitions: large land owners and developers (i.e., real estate interests). Automobiles allow for far-flung land holdings within an urban region to gain utility as housing, manufacturing centers, retail outlets, and office complexes. By doing so, automobiles increase the economic value of such land.

The political and economic relationship between real estate interests and the automobile helps explains why, when confronted with poor air quality arising from the use of the automobile, local growth coalitions opted for technological controls to manage this air pollution and did not seek to restrict or supplant the automobile as the central mode of rapid transportation in urban areas. In the foregoing chapter, Chicago was historically and theoretically significant in terms of analyzing the politics of coal-related smoke. Similarly, Los Angeles takes on historical and theoretical significance

in terms of understanding the politics surrounding (1) the rise of the automobile, and (2) the political effort to control automotive emissions. It was
here in the 1920s where automobile usage was initially routinized or institu-
tionalized within urban planning. This is most cogently explained by Marc A.
Weiss (1987). Also, as I outline in the next chapter, it was in Los Angeles
where air pollution from the automobile first rose to a political level and subsequently where technological controls on automobile emissions were politically pioneered.

This chapter proceeds in the following manner. First, I describe how the
first widely available means of rapid transportation, the trolley, was used to
serve the economic interests of North American real estate concerns. As a
result, North American trolley lines were developed in economically and
geographically inefficient ways during the late nineteenth and early twentieth centuries. This led to economically and politically unviable trolley firms.
It was this lack of viability that facilitated the rise of the automobile and the
demise of the North American trolley system. The second section of this
chapter directly treats how landed interests politically championed the automobile. It was this championing that helped lay the infrastructure (e.g.,
roads, zoning regulations) necessary for the successful operation of the automobile in urban areas.

THE TROLLEY AND URBAN SPRAWL IN NORTH AMERICA

To fully understand how the automobile came to be the predominant means
of transportation in urban North America we must actually go back to an era
before the automobile was available on a mass consumption basis. The sprawl
that created the public's subsequent dependence on the automobile was initiated in the late nineteenth and early twentieth centuries with the introduction of the electrified street railway. The electrified street railway, or the
trolley, in the United States and Canada was utilized more so as a means to
derive wealth from land holdings than as a means to provide efficient and
cost-effective transportation in urban centers.

This is most clearly the case in the Los Angeles metropolitan area. It was
Henry E. Huntington who financed the trolley system that was eventually
deployed throughout the Los Angeles region during the first decade of the
twentieth century (Crump 1988). Huntington had inherited a fortune from
his uncle, Colis Huntington—a co-founder of the Southern Pacific railroad—estimated at $50 million (Crump 1988, 42). He invested much of this
fortune into real estate throughout the region and into the trolley system—
the Pacific Electric and Los Angeles Railway. The system was primarily developed as a way to inflate the value of Huntington's real estate holdings.

Huntington undertook his real estate purchasing efforts in the Los Angeles region during a time when it was a still a fledgling metropolitan area, and

had a relatively small population. Hence, he obtained large tracts of land relatively inexpensively, but these tracts were also widely dispersed throughout the region. Spencer Crump (1988), historian of Huntington's interurban trolley system, the Pacific Electric, in the following describes how Huntington would select where to run his trolley lines and acquire his land holdings:

> Huntington's instinctive business foresight, not a battery of professional economists frequently used by financial tycoons, was his instruments in choosing the areas where his trolleys—and his investments in substantial land holdings—would go. Climbing a knoll, he would inspect the countryside and visualize the logical course for an area's pattern of development. (60)

Utilizing this method, Huntington obtained scattered land holdings throughout southern California. He also deployed a far-flung trolley system that one historian of urban mass transit referred to as "the most extensive interurban [trolley] system in the world" (Foster 1981, 17). This system would disperse economic activity and residential housing throughout the region. This dispersal of development would presage the highly diffuse urban development that would characterize the Los Angeles region throughout the twentieth century and into the contemporary era (Fogelson 1967; Wachs 1984; Hise 1997; Fulton 2001). Crump (1988) argues that when the Los Angeles trolley cars "finally rolled into the realm of history [in 1961], they left a sprawling City of Southern California built precisely as it was because the rail lines had encouraged just that development" (115–116).

The utilization of rapid transit to enhance land values and create urban sprawl was not unique to Los Angeles. The positive relationship between land values and rapid transportation had long been understood (Jackson 1985; Stilgoe 1988). During most of the nineteenth century, however, walking remained the primary means of getting around in urban settings, and as a result cities were relatively compact and often highly congested (Rosen 1986; Schultz 1989).

Early rapid transportation methods were of limited utility. Carriages and omnibuses were reliant on horse power, which severely limited their speed. Moreover, the costs of such modes of transportation restricted their use to affluent city dwellers. The first methods of mechanized urban transportation had limited utility for economic and/or political reasons. The use of steam engines within urban areas was resisted in part because residents feared their explosion. Additionally, the noise and air pollution emitted by such engines tended to depress the value of adjacent property. Also, their long stopping distance, or headway, undermined their usefulness for urban transport. The other early approach to the mechanization of urban transportation was the cable car. Its high initial and maintenance costs, however, confined its use to only the most densely populated areas (McShane 1994).

The trolley did not have the liabilities that did horses, steam engines, or cable cars. Trolleys could move at fairly rapid speeds. Given initial costs and maintenance expenses, trolley systems were relatively inexpensive to run. Trolley cars were also clean and largely noiseless. Thus, soon after their successful demonstration in Richmond, Virginia, in 1887, trolley cars were used in numerous urban areas throughout North America, including New York, Boston, Philadelphia, Chicago, Toronto, and Milwaukee (McShane 1974; 1994; Warner 1978; Davis 1979; Cheape 1980; Foster 1981; Barrett 1983; Fogelson 2001, chap. 2).

Trolley lines throughout North America, like in Los Angeles, were used to increase the value of outlying land. Indeed, the first trolley system in a major metropolitan area was actually developed as part of a massive land development plan (Cheape 1980, 115). Led by Henry M. Whitney, the investors of the West End Land Company sought to develop five million square feet of land on Boston's outskirts. The trolley was the only means to make this subdivided land available to the buyers the company hoped to attract. In order to connect its land holdings with the rest of the city, the investors of the West End Land Company were forced through their subsidiary, the West End Street Railway, to take over the franchises of other transit lines, and subsequently integrated them into a citywide trolley system in the mid-1890s. Historian Charles Cheape (1980) notes with some irony that in Boston "what had begun as an adjunct to a real estate venture became a major transit enterprise" (115). More generally, Mark S. Foster (1981), a historian of rapid transit in North America, notes that "in the late nineteenth century, real estate interests and trolley promoters combined to develop huge areas of Brooklyn, Boston, Chicago, and many other large cities" (17).

Apart from the economic self-interest of trolley owners, another factor precipitating sprawl in urban North America during the turn of the century was political pressure. Clay McShane (1974), for instance, points out that the effort to regulate trolley transportation in Milwaukee ended in the early 1890s when the Milwaukee Street Railway Company agreed to expand into outlying areas (87). In Chicago, a 1907 ordinance mandated numerous "cornfield extensions." They were referred to as such because trolley lines were literally extended into areas made up largely of cornfields (Barrett 1983, 114–115). The Toronto trolley system was taken over in 1921 by local government after the trolley company refused to expand into the city's outskirts (Davis 1979; Weaver 1984). The Detroit railway company was taken over by the city government under the same circumstances (Conot 1986, 186 and 612).

One constant source of pressure for trolley expansion was large land owners and developers.[1] Real estate interests would expend considerable energies lobbying for the extension of trolley lines—sometimes donating land and subsidizing the costs of trolley line construction in order to obtain lines near their holdings (McShane 1974; 1994; Cheape 1980; Jackson 1985, 135; Crump 1988). In Toronto, land developers were the central force behind the public takeover of the Toronto Street Railway and the subsequent expansion of this system into the city's outskirts (Davis 1979, 90; Weaver 1984).

The expansion of trolley systems into sparsely populated areas had a deleterious economic and political affect on them. Other than sprawl, there were factors that sapped the financial strength of North American trolley systems. Key among them was the fact that trolley companies often had to purchase the franchises of competing transit lines in order to reach economies of scale and provide citywide service. The costs of such franchises were in many instances exorbitant. While the costs associated with establishing a citywide system were sometimes high, these and other costs did not undermine the financial viability of North American trolley systems as did their expansion into low population areas (Dewees 1970). Such lines generally lost money, thereby increasing the operation costs of trolley systems. The case of the Pacific Electric and the Los Angeles Railway is telling. While Huntington did not have to expend large sums to purchase competing lines, "financial reports after 1911 . . . showed that the Pacific Electric made a profit for only eight of the forty-two years during which rail passenger service was its business" (Crump 1988, 198). Scott Bottles (1987), who wrote a history on transportation politics in Los Angeles during the early twentieth century, points out that the Los Angeles Railway company, which provided local streetcar service, financially "fared a bit better [than the Pacific Electric], but it never showed the rates of return expected of it" (41). As a result of operation costs, as well as other expenses, North American trolley systems generally lost money or yielded low dividends for its stockholders (Dewees 1970; Davis 1979; Cheape 1980; Foster 1981; Barrett 1983).

The sprawl of trolley systems also created consumer resentment and political difficulties for streetcar firms. Trolley lines running through low density areas put extra cars on the system, thereby increasing traffic congestion. In major metropolitan areas, this congestion could dramatically increase travel time for rush hour commuters. In some cases, so-called short haul commuters found it more efficacious to walk than ride the trolley to work. This, of course, drew down the finances of trolley firms. Another factor fostering hostility toward trolley firms was the general state of their lines. With profits low, or nonexistent, firms were slow to modernize their lines. Thus, many trolley firms employed antiquated, uncomfortable, and sometimes dangerous equipment (Davis 1979; Foster 1981; Barrett 1983; Bottles 1987; Crump 1988). Both trolley line congestion and the general state of trolley systems made efforts to raise fares to offset the costs of sprawl politically contentious (Foster 1981; Barrett 1983; Bottles 1987).[2]

A comparison with the deployment of trolley systems in Europe during the turn of the century serves to elucidate the effect sprawl had on trolley firms in North America. John McKay (1976; 1988) has written the definitive history on the adoption of the trolley in Europe. Unlike the North American case, in Europe trolley lines were generally laid in an economically efficient and geographically effective manner. As a result, European trolley systems did not have the economic and political liabilities that plagued U.S. and Canadian systems.

The political energy and economic resources for the adoption of the trolley in Europe came largely from the manufacturers of trolley equipment. These manufacturers had to overcome the political opposition that developed against the trolley throughout the continent based on aesthetic considerations. The overhead wires that trolley lines utilized led many to object to the use of streetcars on main boulevards. These objections delayed the introduction of trolleys in most European cities for a number of years when compared with the United States and Canada. With the manufacturers of trolley equipment serving as the key impetus underlying the creation of trolley systems, in Europe trolley firms sought to derive a profit from the operation of their lines. Profitable trolley lines could regularly maintain and upgrade their equipment. As outlined above, this is in sharp contrast to the experience of North American cities, where in many important instances trolley systems were created by large land holders and used not necessarily to make a profit from the trolleys themselves but instead from the real estate trolley lines served.

Additionally, in Europe, trolley politics was significantly different than in North America. European governments exerted influence over trolley firms that tended to promote effective and affordable service for patrons. McKay (1976) explains that "it is obvious from the agreements [with trolley firms] that [European] municipalities succeeded in winning substantial improvements in all areas, which went well beyond those inherent in the innovation [of the streetcar], such as speedier, more comfortable, and more hygienic service" (112–113). In the United States, urban governments did not generally regulate the quality of service that consumers received (Foster 1981; Barrett 1983; Bottles 1987). In describing the divergence between the European and U.S. experience with early rapid transit, McKay (1976) makes the point that "a distinguishing characteristic of American tramway development from the beginning was the absence of effective public control" (91). In seeking an explanation for this divergence, McKay posits that "we should . . . note the greater possibilities for the corruption of municipal officials in the United States than in Europe" (94).

With trolley firms seeking profit largely through the operation of their lines, and a political environment that prioritized efficient and effective transportation, streetcar systems in Europe expanded much more conservatively than was generally the case in North America. McKay (1976) argues that European "suburban expansion was facilitated by electric streetcars, which promoted a reasonable diffusion of population without, however, blowing the city apart" (241; also see Ward 1964 and Yago 1984). As a result of cautious expansion strategies, European trolley firms were largely of good financial health.

Moreover, these firms were in better standing with the general public than their U.S. counterparts. In contrast to the European experience, McKay holds that in the United States "free-wheeling entrepreneurial activity, which operated so successfully beyond effective public control, . . . built up strong antipathies" among the general public toward streetcar companies (95). Antipathies that historians Paul Barrett (1983) and Scott Bottles (1987) go to great lengths to document in the cases of Chicago and Los Angeles, respectively.

In the contemporary era, fixed rail transportation persists throughout urban centers in Western and Central Europe. This is facilitated, to varying degrees, through the integration of planning for urban development and planning for mass rapid transport. Additionally, mass rapid transport, including fixed rail, is an attractive option for commuters because European governments assess high excise taxes on the consumption of gasoline.[3] The result of these factors are cities that are generally more compact, and less dependent on the automobile, than those in North America (Hall 1995; Nivola and Crandall 1995; Kenworthy and Laube 1999; Nivola 1999; Beatley 2000; Feitelson and Verhoef 2001).

In the next section, I shift our attention specifically to the rise of the automobile in the United States. The sprawl that occurred in urban North America through the deployment of the electric streetcar during the late nineteenth and early twentieth centuries helped pave the way for the adoption of the automobile during the 1920s in three ways. First, trolley lines moved large portions of the urban populace out of walking distance from city centers. Hence, the sprawl facilitated by trolley systems created a sizable, and largely middle and upper class (Jackson 1985; Fishman 1987), populace dependent on rapid transportation to gain access to goods, services, and employment (Fogelson 2001). Second, the traffic congestion on trolley lines that sprawl helped create in many major cities made the automobile an attractive alternative to the streetcar (Foster 1981; Barrett 1983; Bottles 1987). Third, as the automobile became a serious competitor of trolleys in the 1920s, trolley firms were less able to adjust and respond to this competitor since its finances were undermined by unprofitable lines which ran through low population density zones. Importantly, the finances of most trolley firms made investments in subways or elevated tracks an impossibility (Dewees 1970). These investments could have given streetcars their own right-of-way, free of automobiles and other forms of traffic (Cheape 1980; Derrick 2001; Boschken 2002).

POLITICS AND THE ESTABLISHMENT OF THE AUTOMOBILE

James Flink's (1975; 1990) seminal history of the automobile revolution in the United States emphasizes the technological, organizational, and marketing innovations that facilitated this revolution (also see Curcio 2000 and Farber 2002). Political factors, however, were just as important in the establishment of the automobile as the factors treated by Flink. We saw in the preceding section how the politics surrounding the U.S. trolley helped create a market for the automobile. Just as importantly, it was a political movement that prompted the development of the roads and urban planning that made the automobile a practical reality for urban dwellers. Members of local growth coalitions politically initiated and sponsored the physical and legal changes to the urban milieu that allowed the automobile to operate effectively in this setting.

LAND MANAGEMENT AND SUBURBAN DEVELOPMENT

Marc Weiss (1987), in his instructive work on modern urban planning techniques, explains that such techniques were developed by large land developers in order to protect and enhance their large scale housing developments. Weiss points out that "subdividers who engaged in full-scale community development . . . performed the function of being private planners for American cities and towns." He goes on to write that

> working together with professional engineers, landscape architects, building architects, and other urban designers, residential real estate developers worked out "on the ground" many of the concepts and forms that came to be accepted as good planning. The classification and design of major and minor streets, the superblock and cul-de-sac, planting strips and rolling topography, arrangement of the house on the lot, lot size and shape, set-back lines and lot coverage restrictions, planned separation and relation of multiple uses, design and placement of parks and recreational amenities, ornamentation, easements, underground utilities, and numerous other physical features were first introduced by private developers and later adopted as rules and principles by public planning agencies. (3)

Many of the largest and most innovative land developers "did more than just serve as innovators for the land planning ideas that were spawned in the early 1900s, and spread rapidly during the succeeding four decades." Instead, "many of the large subdivision developers played a direct role in actively supporting and shaping the emerging system of public land planning and land-use regulation" (4). They did so in conjunction with policy-planning groups. Leading examples of such groups include the Home Builders and Subdividers Division and City Planning Committee of the National Association of Real Estate Boards (NAREB). As Weiss explains, most of those large developers, who he refers to as "community builders," involved in shaping government regulations on land use during the early twentieth century "developed stylish and expensive residential subdivisions and were leaders" of this division and committee within NAREB. Significantly, beginning in 1914, a group of community builders from NAREB's City Planning Committee exchanged ideas with the landscape architects, civil engineers, architects and lawyers who predominated in the National Conference of City Planning (NCCP), founded in 1909. Together, these community builders and NCCP activists worked to promote planning legislation among other entrepreneurs, in the real estate industry, to the general public, and within the state and local governments. (Weiss 1987, 56)

Large land developers sought to shape public policies on land use issues because:

> Private developers who scrupulously planned and regulated their own subdivisions needed the planning and regulation of the surrounding private and public land in order to maintain cost efficiencies and transportation accessibility and to ensure a stable and high-quality, long-term environment for their prospective property owners. (4)

A central objective of large developers in championing urban planning during the early twentieth century was to accommodate the automobile. Weiss points out that one of the key factors in prompting large scale community building and the subsequent drive for private and public urban planning was "the increasing availability of private automobiles for upper- and middle-income [home] purchasers" (62). Hence, the accommodation of the automobile in urban and suburban areas, as well as building homes that could accommodate automobiles, became central to reorganizing urban areas and organizing new suburbs (Barrett 1983; Weiss 1987; Hise 1997).

Land developers in the Los Angeles region led the way in the urban planning field. Its sparse population, the fact that developers could purchase large tracts of land relatively inexpensively, and the sprawled out trolley system, meant that Los Angeles was an ideal area to launch large-scale community developments by individual developers. As a result, many of the urban planning techniques and public policies discussed by Weiss were initially developed and applied in Los Angeles (Weiss 1987; Hise 1997).

With Los Angeles developers at the cutting edge of community development methods, they were quick to see the profit potential in the automobile, and planned and developed accordingly. As a result, the mass production of the automobile, and the urban planning methods developed and politically sponsored by large developers to accommodate the automobile, profoundly affected Los Angeles. Foster (1975) points out that

> while the trolley promoters established a number of subdivisions miles from the downtown area, they had developed only a tiny fraction of the land in the Los Angeles area by 1920. Pre–World War [I] residents were so dependent upon the trolley for transportation that developers made few attempts to promote single-family homesites more than a half-mile from the lines. (476)

The declining expense of the automobile and the growing public confidence in it (Flink 1975; 1990), however, "exerted a dramatic effect on the remote areas which were not so well served by the trolleys." Foster explains that:

> The development of the San Fernando Valley during the 1920s was, perhaps, the most spectacular example. The real estate boom of the 1920s witnessed the promotion of thousands of lots, many located miles from the nearest trolley lines. The Encino tract, opened in 1923, contained several hundred single-family homesites. The development was located on the southwest corner of Balboa and Ventura boulevards, two miles from the nearest red [trolley] car stop. The Girard tract—which contained several thousand single-family homesites—was situated even further from the trolley lines, the nearest line being almost three miles distant. These were but two of the many subdivisions opened during the 1920s in the valley where residents generally relied upon the automobile for their transportation. (477)

By the end of the 1920s, the Los Angeles area had become the U.S. region most adapted to the automobile, whereby "residents of Los Angeles purchased more automobiles per capita than did residents of any other city in the country." During this period, "there were two automobiles for every five residents in Los Angeles, compared to one for every four residents in Detroit, the next most 'automobile oriented' American city" (Foster 1975, 483). Historians of Los Angeles take these statistics to assume a particular affinity among the city's residents for the automobile (e.g., Fogelson 1967; Foster 1975; Bottles 1987). A more likely cause, however, for the relatively high level of automobile ownership in Los Angeles is that much of the new affordable housing stock was being constructed in areas only accessible by automobile. Moreover, as businesses responded to the increasing mobility of suburban residents, employment, retail outlets, and services were increasingly offered away from areas serviced by trolleys (Fogelson 1967; 2001; Foster 1975; Hise 1997; 2001). This created further incentives for Los Angeles residents to obtain an automobile.

While the sprawl produced by the combination of the automobile and suburban planning techniques had its earliest manifestation in Los Angeles, other North American cities by the post–World War II period adopted the horizontal development pattern that is reflective of automobile use and large-scale suburban development (Muller 1981; Kenworthy and Laube 1999; Wiewel and Persky 2002). Writing in 1992, Foster observes:

> Many cities, particularly in the Sunbelt, now rival Los Angeles in degree of regional sprawl. Except for foliage, temperature, and humidity, it is often difficult to know if one is in Los Angeles, Houston, Phoenix, or . . . Jacksonville [ellipsis in original]. (191)

One factor that prompted the sprawl of urban development in many U.S. urban areas was the positive opinion that local businesspeople had of horizontal growth. Barrett (1983) documents how in Chicago the business community generally supported the use of the automobile and the outward expansion it brought.[4] Additionally, Blaine A. Brownell (1975), who studied urban public opinion in the South between 1920 and 1930 by examining major newspapers in the region, points out that Southern "businessmen lauded the automobile because it promised to open up new channels of commerce, expand the pool of customers for downtown merchants, and make available large expanses of outlying territory for urban growth and economic development." He adds:

> The major issue concerning businessmen in major southern cities during the 1920s was not whether the automobile was desirable, but whether roads, highways, and related facilities could be provided rapidly enough to insure the maximum degree of economic advantage. The Good Roads Movement in the South, and throughout the country, had always received the support of prominent business groups, and in the 1920s most chambers of commerce in the larger cities established committees especially charged with the task of promoting highway construction and the repair of existing roads. (117)

Howard Preston (1991), who wrote a history on the development of roads in the South during the late nineteenth and early twentieth centuries, adds that "by 1915 the legions of good roads apostles in the South were swollen with chamber of commerce members, bank presidents, sales representatives, real estate agents, and trade board members" (41).

In addition to local support for the automobile and suburban development, beginning in the 1920s, the federal government politically supported horizontal urban growth. Adam Rome (2001), in his book linking the rise of modern environmentalism in the United States to urban sprawl, holds that the federal government during this period fostered low density housing development to attain broad-based home ownership (chap. 1; also see Radford 1996). In accounting for this support of sprawl, Weiss (1987) explains that

> with the accession of Herbert Hoover as secretary of commerce in 1921, NAREB became an important and highly favored trade association working closely with the Commerce Department's new Division of Building and Housing, as well as with other federal agencies. By the early 1930s NAREB was a major presence at the U.S. President's Conference on Home Building and Home Ownership in 1931 and a key national lobbying force behind the creation of the Federal Home Loan Banking System, the Federal Housing Administration, and a number of additional federal policies and programs. (29)

The most significant program undertaken by the federal government to promote home ownership came through the Federal Housing Authority (FHA) (the unofficial name of the Federal Housing Administration). Created in 1934,

FHA's staff was recruited almost entirely from the private sector. Many were corporate executives from a variety of different fields, but real estate and financial backgrounds predominated. For example, Ayers DuBois, who had been a state director of the California Real Estate Association, was an assistant director of FHA's Underwriting Division. Fred Marlow, a well-known Los Angeles subdivider, headed FHA's southern California district office, which led the nation in insuring home mortgages. National figures associated with NAREB, such as real estate economist Ernest Fisher and appraiser Frederick Babcock, directed FHA operations in economics and in underwriting. (Weiss 1987, 146)

As a way to encourage housing sales, the FHA underwrote home purchases. It would guarantee 80 percent of home mortgages for qualified homes and buyers for a twenty-year term. (Later, this guarantee was modified to 90 percent and twenty-five years.) Up to this time, standard mortgages covered 50 percent of the home purchase price and had a three-year term (Weiss 1987, 146).

This program gave the FHA the ability to influence the types of homes purchased and, subsequently, housing development patterns. Weiss notes:

> Because FHA could refuse to insure mortgages on properties due to their location in neighborhoods that were too poorly planned or unprotected and therefore too "high-risk," it definitely behooved most reputable subdividers to conform to FHA standards. This put FHA officials in the enviable position, far more than any regulatory planning agency, of being able to tell subdividers how to develop their land. (148)

With this power, the FHA promoted the building of large-scale housing developments in outlying areas. Weiss (1987) explains that the Federal Housing "Administration's clear preference . . . was to use conditional commitments [for loan guarantees] specifically to encourage large-scale producers of complete new residential subdivisions, or 'neighborhood units'." Thus, the FHA, through its loan program, encouraged and subsidized "privately controlled and coordinated development of whole residential communities of predominately single-family housing on the urban periphery" (147).

Kenneth Jackson (1985), in his important history on the suburbanization of urban development in the United States, concurs with Weiss's assessment of the bias within the FHA for new housing stock in outlying areas. Jackson (1985) writes that "in practice, FHA insurance went to new residential developments on the edges of metropolitan areas, to the neglect of core cities" (206). As a result, Jackson notes that between the years 1942 and 1968 the "FHA had a vast influence on the suburbanization of the United States" (209).

FIXED RAIL TRANSIT IN THE AGE OF THE AUTOMOBILE

As the automobile began to attain widespread usage, North American trolley systems, for the most part, continued to atrophy.[5] Not surprisingly, the automobile worsened the economic position of trolley systems. It did so for two key reasons. First, automobiles took away much needed patronage from trolley lines. Weekend traffic to parks and other recreational venues, for instance, was a substantial source of revenue for trolley firms. Automobiles took much of this traffic, as many purchased an automobile for recreational outings. Second, and more damaging for the trolley, automobiles created more congestion for rush hour trolley riders (Davis 1979; Foster 1981; Barrett 1983; Bottles 1987; Fogelson 2001, chap. 2).

In light of this congestion, the actions and inactions of local and state governments hastened the public's dependence on the automobile and the disappearance of the trolley. Government indirectly exacerbated traffic congestion during the 1920s and 1930s by building networks of roads that could accommodate the automobile (McShane 1988; 1994; Preston 1991; Kay 1998). In contrast, governments generally did not improve fixed-rail urban transportation during this period (Foster 1981).

Efforts during the early part of the twentieth century to subsidize trolley transport, or to give it an advantage over the automobile through the public financing of subways or elevated tracks, were in most instances bogged down in political controversy. A specific national trend was to view subsidized mass transit as an effort by downtown commercial interests to protect the downtown area's position as a center of commerce, during a period when the automobile was making the decentralization of economic activity a reality. Indeed, many proposals that sought to build publicly subsidized subway systems were designed to benefit downtown economic interests, and were politically supported by said interests (Cheape 1980; Foster 1981; Barrett 1983; Derrick 2001; Fogelson 2001). By the 1920s, landed interests in outlying areas often took the lead in politically defeating what were held to be special interest endeavors. Proposed fixed-rail plans that were designed to provide and/or ensure comprehensive citywide service through the usage of public dollars had little political viability during this period (Foster 1981; Barrett 1983; Fogelson 2001, chap. 2). As a result of the political opposition generated by proposals to utilize public funds to improve U.S. trolley systems, historian Robert M. Fogelson (2001) describes how such improvements generally failed to materialize in the United States despite the increasing traffic congestion of urban settings:

> By the late 1920s, after more than two decades of vigorous efforts, after the preparation of scores of studies and reports, after the expenditures of millions of dollars, and after a host of predictions that most big cities would soon build a rapid transit system, nearly 90 percent of the els, close to half of which had

gone up before the turn of the century, were in New York and Chicago. And more than 90 percent of the subways were in New York and Boston, both of which had begun to build their first underground lines before 1900. (109)

While rapid mass transit was mired in political controversy during the 1920s, efforts during this period to expend large sums of public monies to build roads systems in urban areas were largely noncontroversial and undertaken with little or no political friction. Bottles (1987) explains this dichotomy in Los Angeles by arguing that there rapid mass transit was stigmatized with notions of inefficiency and poor service, while the automobile was viewed as a progressive advent that offered freedom of mobility and freedom from the local streetcar companies—the Pacific Electric and the Los Angeles Railway, respectively.

In the case of Chicago, like that of Los Angeles, Barrett (1983) points out that the trolley system was publicly connoted with corruption and monopoly, and this, he holds, contributed greatly to the failure to publicly finance improvements to this system. In contrast, the development of a system of roads for the Chicago area was treated as an apolitical project, and began when the automobile was still a luxury item in the 1910s. The expensive re-development of the Chicago area, so it could accommodate the automobile, was initiated with the *Plan of Chicago,* authored in 1909 (Barrett 1983; McShane 1994, 209–213).[6]

This plan resulted in the creation of the Chicago Plan Commission (CPC). The CPC was dominated by its executive committee, which was comprised "of Chairman Charles H. Wacker, Vice-President-Chairman Francis Bennett . . . , one alderman from each ward, and *a dozen of the city's most prominent businessmen*" [emphasis added]. The long-time managing director of the CPC was Walter Moody. He was also a leading member of the Chicago Association of Commerce, which, as described in the preceding chapter, was an organization composed of prominent Chicago businesspeople (Barrett 1983, 75).

The CPC politically championed the re-organization of Chicago, and received little or no opposition for doing so. Barrett describes how in 1913

> Moody drew a round of applause from an audience of government and business leaders with the declaration that "we ask ourselves too closely 'will it pay?' I sometimes think . . . [that] in other countries . . . they do not so much ask themselves 'will it pay from a monetary standard?' but 'will it pay as an investment, in the happiness and contentment of our people?'" [ellipsis and bracket in original]

And "of a city's debt, when incurred for street widenings and park improvements, Moody argued: 'solvency as it pertains to the City of Chicago does not spell progress; in our case it has spelled retrogression'" (Barrett 1983, 76). Barrett goes on to explain:

Arguments like these, which could never have been made in behalf of mass transit, appear to have done nothing to diminish the standing of the CPC with the business and political elite of Chicago. By 1919 the Commission's managing director [Moody] could refer to it as an "adviser" to the city council, and from the beginning of its existence it rejoiced in the unanimous and enthusiastic support of the press. (76)

The CPC played a key role in the successful effort to reconfigure the political, legal, and physical milieu of the already built-up portions of Chicago so it could accommodate the automobile. First, it was a public champion of the automobile and the means to accommodate it. In part, the CPC undertook this public relations campaign through "hundreds of yearly meetings with neighborhood groups and civic betterment associations" (Barrett 1983, 75). Second, "the Commission found or created the [legal] techniques . . . to help the city adjust to the coming of large numbers of private cars," such as

> the power to spread assessments over a wide area of the south and west sides, and to condemn more property than would actually be used for the improvement, as that part of the land taken could be sold to finance construction. (Barrett 1983, 77)

Finally, according to Barrett, one of the "CPC's most outstanding successes" was "the new North Michigan Avenue," which in the 1920s became "a second CBD [Central Business District], catering to more affluent consumers and relying for access to a large degree upon the automobile and taxi" (78).

While Atlanta was considerably smaller than Chicago, and less wealthy than Los Angeles, it nonetheless displayed a similar road and trolley politics as these other cities during the 1920s. In his history of Atlanta's accommodation of the automobile in the 1920s, Preston (1979) does not note among Atlanta's citizens the hostility toward the local trolley firms that Barrett and Bottles describe in the cities of Chicago and Los Angeles, respectively. Nevertheless, as the automobile grew in usage during the 1920s, no effort was undertaken to publicly salvage the city's trolley system in light of its deteriorating finances. In contrast, the city government of Atlanta and the state government of Georgia expended large sums to create automobile-friendly roads and highways. Significantly, while prominent members of Atlanta's business community, along with the Atlanta Chamber of Commerce, were promoting the automobile and a system of roads and highways, Preston (1979) observes:

> During the twenties the automobile was never challenged as a detriment to the city, and even Preston S. Arkwright, whose street-railway company stood to lose more from the use of motor vehicles in Atlanta than almost any business, never really leveled any sustained attack against the automobile itself.

Preston goes on to observe that "the result of the private use of motorcars and city decision makers' desire for bigness was an Atlanta by 1930 which embraced a 221.31 square-mile area" (150).

While I only highlight three, albeit important, cases here, it appears that the politics surrounding the automobile and trolley during the 1920s and 1930s in Los Angeles, Chicago, and Atlanta were replicated to one degree or another throughout the United States. Foster (1981), in his history documenting the concomitant rise of the automobile and demise of the trolley in the United States, writes that the automobile, on the one hand, won "the minds of most transportation planners" (91), and, on the other, "although there was some concern even in the 1920s over pollution caused by exhaust emissions, there was no crusade against" the automobile (109). Moreover, Foster contends that it was the public's general apathy toward, and ignorance about, the actual financial state of the trolley that lead to the failure of most U.S. cities to move to save urban streetcar systems:

> The 1930s brought the continued decline of the public transit industry in general and the street railway in particular. Obviously, the most important factor behind this decline was the public's preference for individualized mass transit. However, mass transit industry officials must bear some of the blame. Despite growing numbers of pessimists within the industry, too many mass transit executives still deluded the public about the health of America's trolley systems. Partly for this reason, they failed to develop a consensus about the industry's weaknesses and effectively dramatize their requirements. Had either public officials or transportation engineers foreseen mass transit's bleak future, they might have taken more effective steps.

Foster adds that "the country may have lost its last real chance to preserve or develop viable public transportation systems at reasonable cost during the Depression" (131).

CONCLUSION

Politics and public policies played central roles in the rise of the automobile in the United States. These politics and public policies have to a significant degree been historically shaped by large land holders and land developers. Beginning with the trolley, means of rapid transportation were subservient to the interests and preferences of real estate concerns. With the advent of the automobile, many of the larger players in the real estate field quickly seized upon this form of transportation as a means to further maximize the value of their land through low-density development on the urban periphery. These actors, working through such organizations as the NAREB, developed and promoted the private and public planning methods necessary to accommodate the automobile and provide the environs which would attract buyers to the outskirts of metropolitan areas. With the automobile effectively serving the purpose of bringing utility to large swaths of land, the trolley became a superfluous form of urban rapid transit. Hence, publicly financed efforts to preserve or improve trolley systems were viewed as either special interest projects, or as expensive boondoggles.

The resources of local and state governments throughout the United States, as well as of the federal government, were deployed to implement the planning methods developed and promoted by many large scale land developers. As a result, these public entities put forward policies that facilitated and encouraged both suburban development and automobile dependency. These policies included aggressive and expensive road building, and even public subsidies for suburban housing developments.

In Europe, where factors other than real estate values have historically shaped land use and transportation politics, public policies are enacted that encourage mass transit usage and discourage automobile use. Hence, per capita automobile ownership is significantly lower in Europe than in the United States, and vehicle miles traveled per vehicle is also much lower (Hall 1995; Nivola and Crandall 1995; Kenworthy and Laube 1999; Nivola 1999; Beatley 2000; Feitelson and Verhoef 2001).

While certain scholars and policy analysts emphasize the mobility and freedom of movement that the automobile provides (e.g. Bottles 1987; Rajan 1996), the automobile brings with it economic, social, and environmental costs. The one environmental cost we are especially concerned with here is the air pollution that has accompanied the widespread usage of the automobile. The economic and social costs associated with the automobile are also significant. There are the substantial costs associated with the purchase and maintenance of the automobile itself—not to mention the cost of roads, highways, bridges, and their maintenance. Moreover, automobile use consumes a great deal of land—in the form of roads, highways, and parking space. This places upward pressure on the price of land in any given municipality. There are also the millions of individuals that have been killed and maimed as a result of automobile use. Finally, mass automobile utilization has brought large amounts of time lost to traffic congestion. Research has shown that the stress associated with such congestion has negative health effects for drivers (Dreier et al. 2001).

Of these negative effects of mass automobile usage, it is air pollution that adversely impacts local economic growth. With Los Angeles being the first major metropolitan area in the United States to adopt the automobile as the primary mode of local transportation—not surprisingly—it was the first to experience the air pollution problem generated by the mass usage of the automobile. In the next chapter, I describe and analyze how and why Los Angeles responded to this problem.

FIVE

The Establishment of Automobile Emission Standards

AS I NOTED in the last chapter, it was in Los Angeles where air pollution generated by the automobile first reached the political agenda. Subsequently, California was the first jurisdiction in the United States to issue automobile emission standards.[1] In order to fully understand the formulation and limits of California's clean air policies, in particular those that center on the automobile, researchers must focus their analysis on economic elites. Paralleling my findings in chapter 3, central to the effort to regulate air pollution emissions in California are business elites whose economic interests lie in rising property values and an expanding local consumer base (i.e., local growth coalition members).

With such elites providing the key political energy behind the effort to improve air quality in California, especially in the Los Angeles basin, the policy formulation process that established the state's automotive emission standards can be most aptly characterized as a negotiation process between locally oriented economic elites and those national economic forces most closely associated with the automobile industry: automobile manufacturers, oil producers, tire makers, etc.

I begin this chapter by broadly outlining the history of pro-growth efforts in the Los Angeles area. The area's population and economic growth are prime causes of its air pollution problem. The second part of this chapter focuses on the policymaking process that initially established California's automobile emission regulatory regime. This process began in the 1940s and ended in the 1960s. The importance of the early formulation of California's automobile emission regime is that it is here where the contours and trajectory of the current regime were set. These two parts of the chapter demonstrate that many of the same actors, institutions, and economic interests that promoted economic growth in Los Angeles also led the effort to abate air pollution in the area. I conclude the chapter with a discussion of how and why an effort in the late 1960s and early 1970s to challenge the supremacy of the automobile in California failed. This portion of the discussion is drawn from J. Allen Whitt's (1982) instructive work on California transportation politics during this period. In the succeeding chapter, I describe California's current automobile emission regulatory regime, and the politics surrounding this regime.

LOS ANGELES AS GROWTH MACHINE

Just as throughout the frontier West (Robbins 1994; Moehring 2004), many early residents of Los Angeles during the late nineteenth century hoped to profit from growing land values resulting from local investment and local economic growth (Fogelson 1967; Foster 1971, chap. 1; Jaher 1982, chap. 6; Erie 2004, chap. 3). Frederic Jaher (1982) has written an extensive history on the leading political and economic circles of several U.S. cities. In describing the upper socioeconomic strata of Los Angeles, he points out that in this city "the upper class, in common with elites of other eras and places, invested in city lots and rural acreage." Hence, "virtually every leading businessman and politician bought and sold land, and their cumulative influence brought about alienation of the public domain." As a result, "in 1850 the municipality [of Los Angeles] owned 99 percent of the urban acreage but in the twentieth century retained only Perishing Square, Elysian Park, and the Old Plaza" (594). Jaher explains that in Los Angeles "leading realtors typified versatility and kinship, the classic features of the early commercial elite" throughout the United States. He goes on to describe some of the leading commercial elite in early Los Angeles and their business, real estate, and political ventures:

> [Orzo W.] Childs was the initial commercial horticulturalist in Los Angeles, president of the Los Angeles Electric Co., trustee of the Farmers' & Merchants' Bank, a hardware dealer, and streetcar line proprietor. Irish immigrant John G. Downey in 1865 inaugurated the subdividing of the old ranchos. During the 1870s he was the most successful acquisitor of foreclosed estates and built a business block that became the city's new commercial center. Downey arrived in Los Angeles in 1850 and opened a drugstore. He quickly branched out into moneylending, which facilitated his seizures of debt-ridden estates. In the 1850s he served in the state legislature and as lieutenant governor and governor. He established the city's first bank in 1868, built the artesian well in southern California, and financed the second streetcar line in Los Angeles. A railroad director and promoter, he helped the Southern Pacific Railroad gain entry into Los Angeles. J. DeBarth Shorb was associated with his father-in-law Benito Wilson in

extensive speculations, including rancho subdivision, and in citrus fruit growing and marketing. Brothers Joseph P. and Robert Maclay Widney were effective local lobbyists for the Southern Pacific and for harbor improvements, and the latter helped organize the city's initial transit line. The Ohio-born Widneys, nephews of state senator Charles Maclay, another realtor, were respectively a Los Angeles district court judge and president of the Los Angeles County Medical Association, and both helped form the chamber of commerce.

Jaher concludes by pointing out that "the Lankersheim-Van Nuys and Flint-Bixby connections also conducted vast real estate operations" in the Los Angeles area (595).

In an effort to increase access to commerce and attract capital to the area, "several merchants and landowners organized a Committee of Thirty in May 1872" and charged it with inquiring whether the Southern Pacific railroad "could be induced to route its *trunk* line through Los Angeles" (Fogelson 1967, 52 [emphasis in original]). The Southern Pacific, the leading railroad in the state, agreed to run 50 miles of its *"main* track" in Los Angeles County if the county government agreed to cover "5 percent of its assessed value," which amounted to $610,000 (Fogelson 1967, 53 [emphasis in original]).[2]

In light of the railroad's offer, Fogelson (1967), in a history of early Los Angeles, explains that William Hyde, a representative of the Southern Pacific in Los Angeles and of the aforementioned Committee of Thirty, "persuaded the county Board of Supervisors to place before the voters a proposition granting $610,000 to the Southern Pacific for fifty miles of trunk line." Fogelson points out that "at first the bond issue encountered widespread hostility." Apparently, "disillusionment with the railroads was so pervasive in Los Angeles that in the last election each candidate had forthrightly declared his antipathy to rail subsidies." He adds that "several ranchers from southern Los Angeles County, who feared that the donation would increase their taxes, channeled this general dissatisfaction into effective opposition" (53).

"Standing against them," however, "and in favor of the proposition were Los Angeles merchants and landowners who considered a [railroad] connection worth any cost and the Southern Pacific offer their only opportunity" (53). Leading among this pro–Southern Pacific group was Robert M. Widney, "a prominent lawyer and landowner," who published an "influential" pamphlet, entitled "Los Angeles County Subsidy," circulated just prior to the vote on the subsidy (Fogelson 1967, 55). Widney's (1956 [1872]) central point in this pamphlet was that the Southern Pacific proposed connection to the city would greatly increase Los Angeles's trade and population by making it the "second railroad center on the [Pacific] Coast" (358). The specific advantage of having the Southern Pacific run its main line to Los Angeles is that it would connect the area "with the commercial points of the world [especially San Francisco], and does it by more direct routes than any other road can" (351). Fogelson (1967) reports that "by November the pressure from the business community had eroded the antipathy to the subsidies," and the payment to the Southern Pacific was approved (55).

Even before the establishment of the Southern Pacific extension, large landholders sought to attract migrants and capital to the area. Most significantly, "the Southern California Immigration Association, founded by the Los Angeles Board of Trade and supported by prominent property owners, persistently advertised the entire region" (Fogelson 1967, 63). After the Southern Pacific came to Los Angeles, and connected it with San Francisco and New Orleans, this effort started to gain traction. Between 1880 and 1890, the Los Angeles population grew "from 11,183 to 50,395," as did its economy—with the city's "assessed value" growing from "$7 million to $39 million" (Fogelson 1967, 67). This population and economic growth were no mean feats, since numerous areas throughout the country were seeking to attract newcomers and investment. Fogelson explains that Los Angeles's boosters had to compete with "the Great Lakes, Prairie, Rocky Mountain, Southwest, and Pacific Northwest states." Moreover, "nearly all their governments hoped to accelerate settlement and increase property values by channeling the flow of immigration to their regions." These efforts were also "supported by commercial associations seeking to foster trade and encourage industry and assisted by transcontinental railroads trying to stimulate demand for their lands and traffic for their lines" (64–65; also see Robbins 1994).

Despite this intense competition, the Los Angeles area became a vibrant area of population growth and economic activity by the 1920s (Fogelson 1967). The city's railroad connections, which came to include the Santa Fe railroad, and its temperate weather made Los Angeles a leading tourist destination, as well as a magnet to affluent retirees—especially from the upper Midwest (Fogelson 1967, chap. 4). In addition to its natural amenities and infrastructure, the city's growth was spurred by an aggressive publicity campaign undertaken by the Los Angeles Chamber of Commerce. Fogelson (1967) points out that "between 1890 and 1920 the Chamber effectively mobilized the community's resources for promotional enterprises." He goes on to describe these enterprises in some detail:

> [The Los Angeles Chamber of Commerce] established a permanent exhibit of regional agriculture in Los Angeles, encouraged local farmers to participate in fairs and expositions, and shipped their produce to New Orleans, Omaha, Chicago, and San Francisco. More than ten million persons saw these displays of oranges, grapes, and walnuts. The Chamber of Commerce also dispatched a railroad car filled with authentic southern California fruits, vegetables, and spokesmen into rural parts of America.

Another one million people walked through "California on Wheels." Moreover, the Chamber joined with local publishers to distribute Los Angeles newspapers throughout the country, worked with hotel proprietors to attract conventions to southern California, circulated innumerable pamphlets, purchased immeasurable advertising space, and replied to countless queries about the region.

Fogelson holds that "during these years, largely as a result of the Chamber's activities, Los Angeles and environs became the best publicized part of the United States" (70).

By the 1920s, Los Angeles boosters sought to move the city's economy away from its dependency on tourism and its seasonal cycle. In an effort to make tourism to the area a year-round phenomenon, "local businessmen" formed the All Year Club in 1921, which the "Los Angeles County Board of Supervisors divided a yearly appropriation of several hundred thousand dollars between the All Year Club and the Chamber of Commerce for the purpose of promoting Los Angeles' growth during the 1920s." In addition, "the All Year Club's share of the county appropriation was augmented yearly by contributions of $25,000 to $30,000 from the City Council" (Foster 1971, 27).

In addition to the cyclical nature of tourism, Los Angeles economic interests saw perils and disadvantages in relying heavily on tourism for economic growth, because, in part, of their experience during World War I. The war served to bolster the economies of the industrialized cities of the Northeast and upper Midwest, while it undermined tourist-related economic activity. As a result, Los Angeles boosters sought to develop an industrial base for the area (Foster 1971, 24). In addition to the city's experience during the war, Roger Lotchin (1992), a historian of military-industrial investment in California, argues that the desire for industrial economic growth among city leaders grew out of a general belief he terms the "doctrine of industrial advantages." He avers that "the architects of Urban California in particular fervently believed that industrialization held the key to urban stability, continued prosperity, economic diversification, and sectional independence" (5).

In an effort to foster industrial investment in the area, under the auspices of the Los Angeles Chamber of Commerce, the Greater Los Angeles Corporation was created in 1924. Mark Foster (1971), whose dissertation was on Los Angeles's growth and decentralization during the 1920s, explains that

> early enthusiasts envisioned the Greater Los Angeles Corporation as a privately financed organization, which would raise its initial $10 million of operating capital by selling shares of its stock to interested investors for $25 per share. Plans called for an eventual capital base of $50 million, which would be used to assist new local corporations by purchasing blocks of their stocks. The Greater Los Angeles Corporation planned to diversify its

investments and operate just like a mutual fund, presumably selling its shares of a new corporation's stock when it became profitable. It would then be free to reinvest these funds in new local ventures. In such a manner, the corporation would help to boost Los Angeles' industrial growth.

Foster goes on to explain that "nothing developed from the initial organizational meetings" of the Greater Los Angeles Corporation. Nevertheless, its "plans were significant insofar as they revealed the desires of the city boosters to create a comprehensive and fully integrated promotional campaign" involving both national publicity and material aid to prospective industrial investors (27–28).

While the effort to attract industry to Los Angeles through the Greater Los Angeles Corporation came to naught, the city did become a national center of aeronautic investment during the inter-war period (Lotchin 1992; Hise 1997). The actions of local growth advocates played a key role in attracting this investment and making the region a leading manufacturing center of airplanes. Roger Lotchin (1992), in explaining why Los Angeles came to dominate the U.S. aeronautic industry through to 1960, points out that a "critical factor in the seduction of the aircraft industry [to Los Angeles] was . . . the efforts of its promoters to build a great city" (68). In his book, *Fortress California: 1910–1960*, Lotchin describes how, along the lines suggested in the Greater Los Angeles Corporation proposal, the financial, physical, labor, and scientific factors necessary to entice aircraft investment to the region, and to ensure its success, were provided, in large part, by Los Angeles business leaders. Lotchin, in the following, summarizes some of his findings:

> If the plane makers desperately lacked capital, Security Pacific, Brashears, or some private Southland investor provided it when San Francisco or New York City would not. If the industry profited from its proximity to several of the foremost centers of military aviation technology, the boosters had earlier secured these assets. If [plane manufacturers] Douglas, Lockheed, North American, Consolidated, Vultee, and Ryan wanted a cheap and docile labor force, their booster friends did their best to develop and prolong its presence. . . . If the increasingly technological character of the aeronautical industry demanded easy and ever-growing access to both the material and intellectual resources of the scientific community . . . [the California Institute of Technology in Los Angeles] would eventually provide its own airplane testing facilities and then be called upon to manage both the Southern California Cooperative Wind Tunnel and the Jet Propulsion Laboratory. (130)

Lotchin adds that "if an industry largely dependent upon the government for a market [in the form of military contracts] cried out for political influence (and who can doubt that it did?), [southern California] urban politicians from the stature of United States senators down to the city planning commissions stood ready to mobilize it" [parentheses in original] (130).

By the 1940s and 1950s, it was clear that the supporters of the "Greater Los Angeles" project had succeeded in their objective of attracting "the largest possible number of permanent new residents and businesses to Los Angeles from other parts of the country" (Foster 1971, 22). By 1940, Los Angeles County's population had grown almost 200 percent to about 3 million people when compared to a population of about 1 million in 1920, and by 1950 the population of Los Angeles County had reached 5.4 million (Krier and Ursin 1977, 44 and 92). This growth in population coincided with the increasing industrialization of the Los Angeles basin. In addition to expanded airplane manufacturing, growth in Los Angeles during this period took place in such industrial activities as petroleum refining and steel production (Viehe 1981; Boone and Modarres 1999; Hise 2001). As I outlined in the last chapter, this growing population and industry were served by a transportation network centered almost totally on the automobile. Hence, the number of automobiles registered in the county grew from approximately 900,000 in 1930 to 1.2 million in 1940, and by 1950 the total number of automobiles was 2 million (Krier and Ursin 1977, 44 and 92).

With industrialization and the growing automotive population, air pollution came to the Los Angeles area. During the late nineteenth and early twentieth centuries, Los Angeles did not have the severe air pollution problem that plagued urban areas east of the Mississippi in this period because of its comparatively small industrial base and its access to oil and natural gas as sources of energy (Fogelson 1967; Viehe 1981; Williams 1997). By the middle part of the twentieth century, however, industrialization and automobile use, combined with the Los Angeles basin's somewhat unique topography and meteorology, created an air pollution situation that threatened the future growth prospects of the area, and even the gains already made.

"SMOG COMES TO LOS ANGELES"

Historian Marvin Brienes (1976) describes in his article "Smog Comes to Los Angeles" when air quality in Los Angeles became a salient issue. He notes that "on a warm July day in 1943 a mysterious malady settled silently over downtown Los Angeles. . . . [T]he distressing condition worsened daily." This episode "reached its height" on July 26 "as a thick, smoky cloud, heavier by far than any experienced before, descended over the downtown area in the early morning hours and cut visibility to less than three blocks." According to newspaper reports, workers "found the noxious fumes almost unbearable." With this air pollution event Brienes concludes that "smog had come to stay" (515–516). Scholars Krier and Ursin (1977) point to September 8, 1943, as a watershed day in forcing smog onto the political agenda in Los Angeles. On this day,

Los Angeles experienced its "daylight dimout" when dense smog settled over the area. According to one newspaper, "Thousands of eyes smarted, many wept, sneezed and coughed. Throughout the downtown area and into the foothills the fumes spread their irritation."

Everywhere the smog went that day, it left behind a group of irate citizens, each of whom demanded relief. Public complaints reverberated in the press. There was an outraged demand for action. Citizens committees were appointed. Elective officials were petitioned.

They go on to write that "concern over the air pollution episodes of fall 1943, especially that of September 8, stimulated response more substantial than the mayor's [earlier] optimistic prediction of an elimination of smog within a matter of months" (53).

Brienes (1976) outlines how the response of Los Angeles's political leaders to the poor air quality enveloping the area in the early 1940s was to frame the issue in a politically expedient manner. The city's mayor and other city officials publicly blamed a butadiene plant as the cause of the noxious air. Brienes notes that

the fight to tame the butadiene plant had perfectly comprehensible dimensions. It had a beginning, and an end. There was in it an attraction one could never find in contemplation of massive pollution problems that had been woven, over years, into the fabric of the metropolis. (526)

Therefore, by focusing on a single butadiene factory, Los Angeles's political elites were eschewing a more systematic analysis of Los Angeles's air pollution situation—one that might raise politically and economically uncomfortable questions. Brienes explains that "in the butadiene plant a solitary villain was isolated." Hence,

in itself this eliminated a number of potential complications: the need for long, close investigations of thousands of possible offenders; the accumulation of basic scientific data in chemistry and meteorology; planning for permanent, long-range control programs.

As a result, air pollution "abatement became a relatively simple matter of fixing something up or shutting it down" (526–527).

EARLY EFFORTS TO CONTROL AIR POLLUTION
IN SOUTHERN CALIFORNIA

A comprehensive effort to control air pollution in Los Angeles was not undertaken until the Chandler family and its newspaper, the Los Angeles Times, took up the area's air quality matter. The Chandler family throughout the twentieth century had been a central political force in Los Angeles, and, through its newspaper, leading proponents of the area's economic growth (Pincetl 1999, 31 and 96; McDougal 2001). Biographers of the Chandler family and the *Los Angeles Times*, Robert Gottlieb and Irene Wolt (1977), hold that

> [t]he Chandler family, publishers of the Times, has always held a special place as the single most powerful family in Southern California because of its extensive investment and broad political clout in the region. Other newspapers and publishers have played roles in their cities' history, but the extent to which the Chandlers and the Times held sway in Los Angeles was unique. (7)

Another student of Los Angeles history wrote of the *Los Angeles Times* that it was "an enormously influential urban development corporation by itself." In describing Harry Chandler, the individual who established the *Times* as the principal newspaper in southern California, this same historian renders the observation that "it would have been hard to find anyone more closely associated with the booster impulse in Southern California, if for no other reason than that he owned a huge portion of it" (Lotchin 1992, 97).

Harold W. Kennedy (1954), the counsel for Los Angeles County and the County Air Pollution Control District during the 1940s and 1950s, explained that Norman Chandler—by this time the Chandler family patriarch—had "originally sponsored the 'clean air movement' for Los Angeles County" (15; also see Brienes 1975, chap. 5). Chandler's attitude toward the Los Angeles smog question is evident in a statement he made to oil industry executives in 1948:

> The *Los Angeles Times* had entered the [anti-smog] campaign in the public interest with the avowed purpose, if possible, of finding all the sources of air pollution, and was committed to the position of going forward without fear or favor irrespective of its effect upon any industry. (paraphrased in Kennedy 1954, 15)

In an effort to build a political consensus on the issue of air pollution abatement in Los Angeles, "Chandler recruited a new citizens' committee" in 1946 (Brienes 1975, 123). Marvin Brienes (1975), who wrote his dissertation on the effort to abate smog in Los Angeles between the years 1943 and 1957, notes:

> Known at first as the *Los Angeles Times* Citizens Smog Advisory Committee, the new group boasted a prestigious membership, including Dr. Robert A. Millikan of the California Institute of Technology, the Rotary Club president, Don Thomas of the tourism-boostering All-Year-Club, and [Stephen W.] Royce [owner] of the Huntington [Hotel]. (123)

Brienes adds that "at the first working meeting" of the committee, "January 1947, [William] Jeffers [official head of the Citizens' Smog Advisory Committee] said 'a great deal of missionary work' lay ahead [in convincing others of the need for government action to reduce air pollution in Los Angeles], and announced that he would begin his with the Los Angeles Chamber of Commerce" (123).

In 1946 the *Los Angeles Times* brought in Prof. Raymond Tucker from Washington University in St. Louis to study and report on the Los Angeles air pollution situation ("'Times' Expert Offers Smog Plan" 1947; Brienes 1975, chap. 5). Tucker was smoke commissioner and in charge of regulating industrial air pollution emissions in St. Louis from 1937 to 1942 (Ainsworth 1946). His report analyzing the sources of Los Angeles's smog was completed in January 1947 and published on the front page of the *Times* (Kennedy 1954, 5; Air Pollution Foundation 1961, 6; Brienes 1975, 123–125; Ursin and Krier 1977, 57–58). Tucker, in his report, specifically pointed to the "chemical industries, refineries, food products plants, soap plants, paint plants, building materials, nonferrous reduction refining and smelting plants, as well as numerous others of similar types" as major sources of air pollution in the Los Angeles basin, and he advised government action to regulate their airborne emissions ("Text of Report" 1947).

Tucker's key recommendation was that a single countywide district be created to regulate air emissions in the area. He specifically urged that "the necessary State legislation be enacted to create an air pollution control district, preferably county-wide" ("Text of Report" 1947). A countywide pollution district would overcome the difficulties of having to enact and enforce regulations throughout Los Angeles County's numerous municipalities and unincorporated areas. Tucker's recommendation was the central feature of air pollution legislation introduced in the California legislature in early 1947 ("'Times' Expert Offers Smog Plan" 1947; Brienes 1975, chap. 5).

Los Angeles city Mayor Fletcher Bowron sought to alter the proposed legislation by backing "a plan to give incorporated cities a two-thirds majority on the governing board of the smog district" (Brienes 1975, 126). Bowron's position was shared by certain Los Angeles county "municipalities, or their agencies, fearing loss of autonomy" (Brienes 1975, 125). After "a series of meetings" with a legal committee created by the *Los Angeles Times* Citizens Smog Advisory Committee, and "headed by James L. Beebe, a former head of the Los Angeles Chamber of Commerce," Mayor Bowron and the "objecting cities abandoned their opposition" (Brienes 1975, 125–126).

Significant support for the legislative effort to a create countywide smog control agency came from the Los Angeles business community. Business organizations supporting the bill included the Automobile Club of Southern California, the Los Angeles Chamber of Commerce, and the Pasadena Chamber of Commerce (Kennedy 1954, 14). The "editorial department of the *Los Angeles Times* . . . [was] directed by Norman Chandler, publisher, to vigorously support the smog legislation." Kennedy goes on to report that the editorial board put forward a "newsletter [in support of the legislation] which was published in the evening edition of the *Los Angeles Times* on May 17, 1947 followed by a strong editorial in the Sunday *Times* of May 18, 1947" (Kennedy 1954, 13). These pieces in the *Times* were particularly critical of the oil refining industry, and its opposition to the proposed clean air legislation, which was preventing its passage through the state senate (Ainsworth 1947; "Public Called on to Block Crippling of Anti-Smog Bill" 1947; Kennedy 1954, 11; Brienes 1975, 129–130).

Kennedy, as counsel for the County, was invited to a meeting of oil company executives, held on May 19, 1947, where the executives discussed the political position of their companies on the proposed smog legislation. Attending the meeting were representatives from Standard, Union, Texaco, General Petroleum, Shell, and Richfield. Also present was William Jeffers, who in addition to heading the *Los Angeles Times*'s Citizens' Smog Advisory Committee, was a former chairperson and president of the Union Pacific Railroad Company ("'Times' Expert Offers Smog Plan" 1947; Kennedy 1954, 13–14).

Kennedy (1954) explains how oil companies during this meeting came to withdraw their initial opposition to the smog abatement bill. Central to this decision was the support that some company executives expressed for the legislation:

> At this meeting the question of the official position that should be taken by the major oil companies was fully discussed by the executives of these companies. The leadership of Mr. William Stewart, Vice President of the Union Oil Company, was invaluable. He argued that smog was a far reaching community question affecting not only the comfort and health of the entire community, but also *its future prosperity*. Mr. Stewart spoke for Union Oil Company and firmly stated that as far as his company was concerned, they believed they must make their contribution to Los Angeles County, and to Southern California and for their part they would not oppose the passage of Assembly Bill 1 [the anti-smog bill]. In the presence of Mr. Jeffers, and Mr. Kennedy, a frank discussion was held and at the conclusion Mr. Charles Jones, President of Richfield, who had called the meeting and served as Chairman, went around the table and the roll informally was called, indicating the withdrawal of the theretofore expressed opposition of the major oil companies [emphasis added]. (14)

Shortly after the oil company executives reached their consensus, Kennedy reported that he received a telephone call from a leading oil industry lobbyist explaining "that the oil companies would not further oppose the passage of Assembly Bill 1 and would not insist upon any amendments to the bill" (Kennedy 1954, 14). Just prior to the oil companies' meeting, Jeffers, in a meeting with the major railroads operating in California, convinced these firms to withdraw their opposition to the smog bill (Kennedy 1954, 10; Brienes 1975, 128–129).

The bill was approved in the state Senate by a vote of twenty-nine to zero and was promptly signed by the Governor. It led to the creation of the Los Angeles Air Pollution Control District (APCD), established with a budget of $178,000. As one historian of U.S. air pollution politics explains, this made it the "best funded air pollution control agency in the nation" (Dewey 2000, 44). Between the years 1948 and 1957 the APCD expended nearly $10 million on pollution control efforts (Dewey 2000, 57).

THE INITIATION AND FORMULATION OF CALIFORNIA'S AUTOMOBILE EMISSION REGIME

Despite the work of the APCD (Brienes 1975, chaps. 6 and 7), air pollution continued to be a persistent, if not worsening, problem in the late 1940s and early 1950s (Air Pollution Foundation 1961, 7; Carlin and Kocher 1971, 57; Krier and Ursin 1977, 72). A particularly egregious five days of heavy smog in Los Angeles during the fall of 1953, known as the "five-day siege of smog," prompted locally oriented economic elites and the automotive related industries to create a policy-planning organization, the Air Pollution Foundation (Air Pollution Foundation 1961, 8; Krier and Ursin 1977, 83). The Foundation board of trustees was composed almost entirely of representatives from business and industry. Many of these individuals represented firms whose future, in part, depended upon continued economic growth in Los Angeles. Among these locally oriented firms were the Southern California Gas Company, Bank of America, Broadway-Hale Stores (department stores), Western Air Lines, California Federal Savings, California Bank, Southern California Edison Company, Security-First National Bank, and Bullock's (department stores). Among the automotive-related firms represented on the Foundation's board were each of the Big Three automakers, Firestone Tire & Rubber Company, U.S. Steel, and the Union Oil Company (Air Pollution Foundation 1961, 50–51). The Foundation's list of contributors demonstrates the broad support that it enjoyed throughout the corporate community. Throughout its seven-year existence (1954–1961) the Foundation had more than 200 donors, almost all of which were from corporate America. Among its financial supporters were the Automobile Manufacturers Association, the Western Oil and Gas Association, the Los Angeles Newspaper Publishers Association, the Los Angeles Clearing House Association (made-up of major California banks), DuPont, Bechtel Corp. (construction), Kaiser Steel, and the Goodyear Tire and Rubber Company (Air Pollution Foundation 1961, 53–56).

The Foundation sought "to assemble a competent technical staff to organize and direct a broad program of cooperation, research, and public information" on the issue of smog in southern California. Additionally, the Foundation planned to "determine what remains to be done and to employ

experts—through the device of research or service contracts—who will provide information and advice for the shaping of future policies and action"
(Air Pollution Foundation 1961, 8–9). Therefore, the Foundation, and the
economic elites that composed its leadership, hoped to collect information
and technical analysis to develop public policies to manage the Los Angeles
smog situation.

Throughout its existence, the Foundation examined studies and sponsored its own research with regard to the issue of air pollution. A particular
focus of its investigation centered on the automobile. Here the Foundation
considered studies put forward by such organizations as the Automobile Manufacturers Association, the Franklin Institute (for the American Petroleum
Institute), the APCD, the University of California, and the University of
Southern California (Air Pollution Foundation 1961, 22). The Foundation
also conducted its own research on the role of the automobile in the formation of smog in Los Angeles.

As a result of this work, by the end of 1956 "it became apparent to the
Foundation that motor vehicles were the principal contributors to smog in
Los Angeles" (Air Pollution Foundation 1961, 25). Significantly, shortly
after the Foundation reached this conclusion, Krier and Ursin (1977) point
out that "consensus on this point grew quickly" (86). In particular, the automobile industry dropped its longtime position that the automobile was not a
major contributor to the formation of smog in Los Angeles (Campbell 1953;
Ford 1953; Chayne 1954; Krier and Ursin 1977, 89). Soon after this consensus was reached, California in 1960 enacted legislation requiring the installation of pollution control technology in automobiles (Krier and Ursin 1977,
chap. 10; Dewey 2000, chap. 4).

After the Foundation determined that the automobile was a major
source of smog-causing pollutants, its leadership then decided "that the future
program of the Foundation would be directed almost completely to a study of
motor vehicle exhaust and its control" (Air Pollution Foundation 1961, 26).
In terms of the control of automobile exhaust, the Foundation's answer was
technology. It centered its "research program" on the "development of scientific principles upon which effective exhaust control devices could be used"
(Air Pollution Foundation 1961, 29). By relying exclusively on technology to
address the smog derived from automobile emissions, the Foundation also
defined the problem in a way that served the economic and political interests
of its membership and donors. By proffering technology as the sole solution
to automobile emissions, the problem is then defined or framed (Baumgartner and Jones 1993; Laird 2001) as a lack of effective emission control technology, and not a problem caused by too much economic and population
growth or too many automobiles in the Los Angeles basin.

When automobile emission standards were established on a statewide
basis in 1967, it was under the guidance of Dr. Arie J. Haagen-Smit, who

served as the longtime chairperson of California's Air Resources Board (Krier and Ursin 1977, chap. 11). Haagen-Smit was a member of the Foundation's technical board, and prior to that was on the Los Angeles Chamber of Commerce's scientific committee (Air Pollution Foundation 1961, 52; Brienes 1975, 90–92; Krier and Ursin 1977, 79).

THE PUBLIC AND CALIFORNIA POLLUTION ABATEMENT POLICIES

Writing about pollution control policies in Los Angeles during the 1940s, 1950s, and 1960s, Thomas Roberts (1969), in his well documented and researched honors thesis, explains that during this period "in general the mass public . . . remained politically quiescent with regard to smog control" (46). During this time no mass-based organization came into being that could organize and mobilize the public on the issue of air pollution (Roberts 1969, chap. 3; Krier and Ursin 1977, 272–277). The three citizens' groups pushing for pollution controls that Roberts identifies all had a small number of members.[3] Further, the one that Roberts considers to be "most important" could be said to represent the economic elite perspective on smog. This economic elite organization, known as Stamp Out Smog (SOS), was composed of Beverly Hills housewives (Roberts 1969, 48–52; Dewey 2000, 97–98). Significantly, SOS members did "not seriously consider staging the consumer boycott of new autos which many believe[d] ultimately necessary" (Roberts 1969, 51). Instead, it focused its efforts on prompting the automobile industry to develop and distribute emission control technology (Roberts 1969, 49).

The other two groups noted by Roberts were the Group Against Smog Pollution (GASP) and the Clean Air Council (CAC). GASP was made up of faculty members from the Claremont Colleges, and CAC was composed of "scientists, engineers and other professionals" (Roberts 1969, 51). GASP tried unsuccessfully to organize a boycott of new automobiles (Roberts 1969, 51). CAC "drafted" a voter initiative in the late 1960s that would require "strict, explicit, emissions standards, and in effect forcing off the road any car which is not pollution free." Furthermore, it developed plans "to implement rapid transit" and "alternative propulsion systems to the internal-combustion engine" (Roberts 1969, 51–52).

Pluralists (e.g., Dahl 1961; Bryner 1995) and other thinkers (e.g., Rajan 1996; Mazmanian 1999), hold that policy development—to one degree or another—is a product of the relationship between public officials and public opinion. In this case, they would argue that while the public was, for the most part, unorganized and politically immobile on the issue of air pollution in Los Angeles during the 1940s, 1950s, and 1960s, the enactment of air pollution legislation and the creation of regulatory bodies during this period were nonetheless in response to the public's latent opinion on air pollution. Thus, according to the thinking of these researchers, the potential mobilization of the public was enough to prompt substantial policy action to address the Los Angeles smog problem.

In contrast, critical theorists (e.g., Cahn 1995; Aronowitz and Bratsis 2002), and especially those that center their analyses on economic elites (e.g., Weinstein 1968; Miliband 1969; Domhoff 2002), would hold that the legislation of the 1940s, 1950s, and 1960s, and the regulatory agencies that resulted from this legislation, were in part the result of efforts to keep public opinion on the issue of air pollution latent. Empirically, the policy formulation process described above is consistent with the critical approach to reform politics. One can assume that the economic elites that promoted emission controls in Los Angeles did so in part to prevent the public from mobilizing around this issue and seeking to impose their own, and potentially radical, solutions to air pollution, either through legal means, such as voting, or extra legal means, such as mass demonstrations (Ford 2001). The existence of public policies utilizing technology to abate air pollution communicated to the broader public that it need not try to mobilize on the issue of air quality because substantive actions were already addressing the area's acute smog problem (Cahn 1995).

The work of the Air Pollution Foundation does demonstrate that economic elites sought to shape public opinion and assuage it on the issue of air quality. In describing the milieu of the Foundation's origins, the authors of its *Final Report* point to what was perceived as increasingly negative public opinion with regard to the issue of air quality in Los Angeles. They explained that in the late 1940s and early 1950s "public disappointment" over air pollution abatement policies "grew into intensive criticism of the APCD, fanned by speeches and 'Letters to the Editor' by the overzealous, by fanatics, and even by well-meaning citizens who were led to exasperation by the turn of events" (Air Pollution Foundation 1961, 7). One of the Foundation's primary goals was to "publish current information—by the most appropriate means—on all phases of air pollution and its abatement" (Air Pollution Foundation 1961, 8–9).

Also, after what was deemed by the Foundation (1961) in 1955 as "the worst smog in Los Angeles' history," it concluded that "the need for a sound public information program was obvious." As part of this "public information program" the Foundation "published 10 technical reports . . . and distributed them widely." Furthermore,

> a monthly newsletter was initiated. Two speakers' bureaus, . . . one formed by business and industrial supporters of the Foundation, were kept active. Sound motion pictures and color slide collections were made available to these groups as visual aids. The annual meeting of the Foundation was held in conjunction with the Southern California Conference on the Elimination of Air Pollution arranged by the California State Chamber of Commerce in cooperation with the APCD and the Foundation. (21)

As a way to shape the views of high school students on the issue of smog, "a 20-page pamphlet titled 'Air Pollution and Smog' was prepared by . . . the Foundation Public Information Officer. . . . Approximately 18,000 copies were distributed by the Foundation and by the Los Angeles County APCD." Further, according to the Foundation, its newsletter served the function of "keeping opinion leaders apprized of current air pollution problems and developments in Los Angeles" (44). Through these public information activities, the Foundation, specifically, and the business community, generally, promoted its solution of technology as the only appropriate means to address air pollution in Los Angeles (Air Pollution Foundation 1961). In his 1959 annual message to the Air Pollution Foundation Board of Trustees, Foundation President Fred D. Fagg Jr. declared that

> through its public information activities, the Foundation has won the confidence of opinion leaders and important support for the thesis that *given time and the necessary funds, science and engineering will eliminate smog* [emphasis in original]. (Air Pollution Foundation 1961, 40)

TRANSPORTATION POLITICS IN CALIFORNIA IN THE 1960s AND 1970s

While the leadership of the Air Pollution Foundation sought to limit the automobile pollution debate in California to one of pollution control technology, others sought to expand the discussion over transportation to include mass transportation and land use during the late 1960s and 1970s (Fellmeth 1973; DeLeon 1992; Carter 2001; Doherty 2002). The most significant effort to provide public funding for mass transportation in California occurred with Proposition 18, which was placed on the November 1970 ballot.

In 1962 voters did approve a $792 million bond issue to finance the Bay Area Rapid Transit (BART) system. As Whitt (1982) explains, however, "BART is not designed to challenge the dominance of the private automobile in the Bay Area." Instead,

> BART was designed to serve other goals, goals that are not in conflict with the continuation of automobile dominance. Essentially, these goals were the preservation and growth of the central city and the protection of corporate investments there.

Furthermore, "the prime initiators and supporters of BART were the giant corporations located in downtown San Francisco." Moreover, "there was very little involvement by citizens' groups and there was no opposition to BART by California's famous highway lobby because it was realized that BART was a supplement to the private automobile, not a replacement for it." Finally, BART "was to be financed out of bonds and property taxes, not out of the highway trust fund" (41).

Thus, within thirty years of numerous automotive-related companies, among them General Motors, Standard Oil of California, and Firestone Tire and Rubber, being found by a federal grand jury to have successfully conspired to dismantle electric streetcar systems in forty-five U.S. cities, including Los Angeles, San Francisco, and New York (Snell 1974; Yago 1984, chap. 4; Bottles 1987), a major California city was financing the creation of a fixed-rail system. As outlined by Whitt, however, this rail system was specifically initiated and designed to relieve congestion in the central city district, and facilitate travel to it. It was not a comprehensive effort to provide a citywide publicly financed rail system—one that could free many from the automobile.

Proposition 18, however, would have made a substantial amount of money available for the development and operation of mass transportation throughout the state. This proposition would have allowed the state legislature to use an unspecified amount of highway trust fund dollars for the purpose of "control of environmental pollution caused by motor vehicles" (as quoted in Whitt 1982, 105). Additionally, "the proposition would have given voters in local areas the option of using up to 25 percent of the [the highway trust fund] revenues collected in their city or county for mass transit purposes." Whitt (1982) aptly notes that "Proposition 18 was a mild measure: local voters could elect to continue using all local funds for streets and roads, or could use a maximum of one-fourth of those funds for mass transit" (105). Nevertheless, the proposition created the potentiality of having significant funds diverted from the highway trust fund toward the building and operation of fixed-rail transportation. The highway trust fund was created in 1938, and is financed through a gasoline tax and automobile licensing fees. Its funds were constitutionally designated for highway construction and maintenance.

Once Proposition 18 received two-thirds of the legislature's approval as required for a state constitutional amendment, it was placed on the ballot. Unlike the BART bond issue, the campaign in favor of Proposition 18 "was organized by civic and environmental groups." Specifically, "the main driving forces behind the organization of [the pro-proposition] campaign were the Sierra Club and TARDAC [the Tuberculosis and Respiratory Diseases Association of California]." In addition, "the Coalition for Clean Air, representing various environmental groups in Southern California, provided volunteer labor" (Whitt 1982, 107).

While a number of organizations expressed support for Proposition 18, including the League of Women Voters, the California Medical Association, League of California Cities, and Californians Against Smog, "virtually all businesses that announced a position were opposed." Joining the Automobile Club of Southern California and numerous major oil companies in opposition, were the California State Chamber of Commerce, and the California Real Estate Association (Whitt 1982, 111–112). Notably, the Los Angeles Chamber of Commerce expressed support for the proposition, but "proponents contend that the endorsement was half-hearted and came too late to aid their cause" (Whitt 1982, 117).

Whitt (1982), in the following, outlines the spending by both sides on the Proposition 18 campaign:

> Opponents spen[t] about fifteen times as much as proponents ($333,455.69 versus $22,721.81). Total contributions against and for 18 were $348,830.00 and $17,714.20 respectively. The big money was overwhelmingly on the side of the opposition. Also, it was almost entirely (98.6 percent) business money in opposition (i.e., from the highway lobby). (122)

This anti-18 campaign funds total is probably a significant underestimation of opponents' total expenditures. Robert Engler (1961; 1977), in discussing the political activity of oil firms, in particular, explains that campaign reporting rules in the 1960s did not gauge the numerous ways this industry would give to political campaigns. He (1961) specifically points to the fact that extant campaign disclosure practices did not collect information about the "campaign offerings that are concealed and unreported through cash giving, misleading listings, padded expense accounts, and dummy bonuses for executives, with the understanding that the money is to go for politics." Also, Engler correctly asserts that "institutional advertising, 'educational' and association activities are often essentially political, as are the quiet loan of corporate facilities and personnel" (366; also see Olien and Olien 2000).

Engler's insights into the political practices of the petroleum industry are pertinent here because oil companies "played the predominant part in opposing Proposition 18 not only in terms of money [75.1 percent of the reported total], but also in terms of organization." Specifically, "Harry Morrison, general manager of the Western Oil and Gas Association, and ex-public relations officer for Shell, Carl Totten, . . . worked on the opposition media campaign" (Whitt 1982, 124). With the anti-18 campaigners able to finance and man an aggressive media "'blitz' via television for the last seven days, newspapers for the last five days, and radio for the last five days," the proposed constitutional amendment was defeated by a vote of 2.7 million (45.9 percent) to 3.2 million (54.1 percent) (Whitt 1982, 121).

In 1974 California voters did approve Proposition 5 (Whitt 1982, 129–132). This amendment to the state constitution allowed governments within California to ultimately divert 25 percent of their highway trust monies toward mass transit, much like Proposition 18 in 1970 proposed to do. This approved amendment allowed the state to take advantage of federal matching funds for mass transit projects. Significantly, unlike Proposition 18, Proposition 5 only allowed the diversion of highway trust dollars for new mass transit capital outlays, and prohibits any expenditures for the maintenance or operation of mass transit systems. Such funds would have to be raised through other revenue mechanisms (e.g., fares [i.e., user fees], sales, income, or property taxes).

A number of major corporations supported the 1974 proposition, including Bank of America, Atlantic-Richfield (an oil firm), and Crocker National Bank.[4] Importantly, major oil firms did not come out in opposition to Proposition 5 (Whitt 1982, 131). Detailing the lack of any serious opposition to this proposition, Whitt (1982) explains that "only $1,700.29 was put up against Proposition 5." He adds that "$203,215 was contributed" to support passage of the measure, "with 99.4 percent coming from business" (132).

CONCLUSION

In examining the history of Los Angeles's economic growth and the creation of California's air pollution abatement regime, we can see that many of the individuals, institutions, and interests that promoted and economically benefitted from Los Angeles's growth also took the lead in shaping and establishing California's pollution control efforts. Their air pollution reduction and growth objectives were reconciled through the utilization of pollution control technology to achieve the former. Hence, economic growth could continue and air quality could improve. Improved air quality would, in turn, protect the business milieu.

The technology control approach to air pollution abatement was extended to automobile emissions. A conciliatory attitude was taken toward the automobile, not solely because of formal economic ties among automotive-related firms and leading economic interests in the state (Whitt 1982), but, as I discussed in chapter 4, large land holders and land developers in Los Angeles, and throughout the state, by the 1940s and 1950s had become as economically dependent on the sale and use of automobiles as automotive interests themselves. Thus, when the opportunity opened to shift a significant amount of financial resources away from the development and maintenance of publicly financed automotive infrastructure and toward the development, maintenance, and operation of non-polluting mass transit, local growth coalition members in the state stood largely idle while automotive-related interests, especially oil, closed it off.

Finally, the policymaking process that led to the establishment of the automobile as a major contributor to smog, and the subsequent creation of California's automobile emission regulatory regime, demonstrates that the current policy approach taken by public agencies in California to the issue of automobile emissions is not the result of public officials seeking to navigate the competing and somewhat conflicting preferences of the public, as posited by Sudhir Rajan (1996). Nor is this approach the result of public officials striving to reconcile economic growth and air pollution concerns, as held by Daniel Mazmanian (1999). Both Rajan's and Mazmanian's arguments are consistent with the state autonomy/issue networks model. Instead, we see that the current contours of California's clean air policies resulted from the coordinated efforts of economic elites to develop an approach to reducing air pollution that would protect their collective economic and political interests.

California remains as the U.S. center of policymaking on the matter of automobile emission standards. Massachusetts, for instance, by law ties its standards to those of California (Ridge 1994). By the 1970s, environmental groups were seeking to directly shape California's public policies on air pollution abatement, especially as they related to the automobile. How have these efforts affected the state's and nation's clean air politics and policies? It is to these issues that I turn to next.

SIX

Democratic Ethics, Environmental Groups, and Symbolic Inclusion

THE STATE OF CALIFORNIA is the nation's leader in the formulation and implementation of automobile emission standards. Its automobile emission standards are the toughest in the United States. These standards were most recently tightened in 1998. Additionally, as I explained in chapter 1, California's standards have been the driving force behind the nation's automobile emission standards. Specifically, the strengthening of California's emission standards in 1990, in conjunction with the actions of other states, prompted the federal government to raise its automobile emission standards with the 1990 Clean Air Act. Similarly, events at the state level, led by California, prompted the federal government in 1999 to announce a tightening in emission standards (Cone 1998; Perez-Pena 1999).

Despite California's well-developed regulatory framework and its political leadership on the issue of automobile emission standards, air pollution from automobiles continues to be a persistent and serious problem in the state. While carbon monoxide and nitrogen oxide emissions in California are down when compared to 1990 levels, these pollutants continue to be emitted in large and hazardous amounts into the state's air. Additionally, the amount of particular matter in California's air remain at roughly 1990 levels. Moreover, particulate matter is predicted to increase in the near future. The automobile (including buses and trucks) accounts for at least 80 percent of all these pollutants in California's air. Further, automobiles are the primary source of such airborne toxins as acetaldehyde, benzene, 1,3–butadiene, formaldehyde, and diesel particulate matter (known to be a carcinogen) (California Air Resources Board 2004). Largely as a result of automobile usage (California Air Resources Board 2004, 60–61), Los Angeles, and the region surrounding it, continues to have the most ozone (i.e., smog) polluted air in the United States (American Lung Association 2004).

If California has the most stringent automobile emission standards in the country, indeed the world, why does air pollution from the automobile continue to be a persistent problem and a potentially increasing problem in the future? The primary reasons are population growth, growing economic activity, and an increase in the number of automobiles (with internal combustion engines) as well as in the average number of miles driven (Kenworthy and Laube 1999; Lange 1999; Patton 1999; California Air Resources Board 2004). Sacramento's population, for example, is expected to grow by 50 percent in 2010 from 1987 levels, and it is expected to have an increase of 76 percent in the number of miles driven (Grant 1996, 34). Therefore, the gains in emission reductions made through the application of technology to the internal combustion engine are offset by the overall increase in the number of automobiles on the road and an increase in the average number of miles driven by motorists (Kamieniecki and Farrell 1991; Warrick 1997; Cone 1999; Luger 2000; Purdum 2000). Hence, despite the increasingly onerous regulatory framework placed on the automobile in California, the state, especially the Los Angeles basin and the Central Valley area, will continue to have unhealthful air into the foreseeable future (California Air Resources Board 2004).

Given the relationship between growth, automobile usage, and air pollution in California, is the regulation of growth and automobile usage actively considered in order to reduce and remedy the considerable air pollution problem in the state? The answer is no. In his study of California's automobile and fuel emission standards, Grant (1996) analyzes the "policy community" surrounding this issue area.[1] Borrowing from Rhodes and Marsh (1992), he describes a policy community as "characterized by a limited number of participants, frequent interaction, continuity, value consensus, resource dependence, positive sum games, and regulation of members" (Grant 1996, 10). In his analysis of the California air pollution policy community, Grant (1996) concludes that issues of land management and mass transit are excluded from it. In contrast, technological solutions are at the center of the clean air policy community.

With the issue of growth and the number and usage of automobiles excluded from government's clean air agenda (Bachrach and Baratz 1962; Lukes 1974), the goal of its automotive emissions regulatory regime can be characterized as the ecological modernization of the automobile. Moreover, with the mass production and distribution of alternative fuel automobiles becoming less likely (Borenstein 2000; Pollack 2000; Steele and Heinzel 2001; Roberts 2004), the objective of this regime can in retrospect be termed as the ecological modernization of the gasoline-burning internal-combustion engine.

At the core of ecological modernization is the idea that environmental protection and economic growth are complementary goals. This complementary relationship can be achieved primarily through the development and application of technology. According to its proponents, the costs associated with ecological modernization are justified because an ecologically modernized society produces less pollution and hence utilizes materials more efficiently. Further, advocates of ecological modernization hold that a cleaner environment results in greater productivity. Additionally, the ecological modernizing of consumer products leads to economic growth and increases competitiveness because consumers are increasingly demanding environmentally benign products (Weale 1992; Hajer 1995; Dryzek 1996a, 480; 1997, chap. 8; Mol 2002; York and Rosa 2003).

This approach is consistent with free market environmentalism, which holds that environmental protection is consistent with the utilization of the market (Dryzek 1997, chap. 6; Anderson and Leal 2001). The difference between the advocates of ecological modernization and the free market environmentalists is that the modernization school holds that market mechanisms cannot be solely relied upon to achieve a salutary environment. Instead, public regulations are often necessary to correct for market failures and advance the ecological modernization of capitalist society.

With the objective of California's, and the nation's, automotive emission regime being the ecological modernization of the internal-combustion engine, the continued participation of environmental activists in the policy process that formulates this regime raises ethical dilemmas. These ethical dilemmas flow from a normative framework of democracy. Specifically, the participation of environmentalists within the policymaking process that formulates California's, and subsequently the country's, automobile emission regulatory regime serves to undermine the potential development of a broad-based movement that could debate and challenge the narrow contours of this regime. Additionally, this narrow regime is not driven by those environmentalists incorporated into the policymaking process, as is often argued (e.g., Sabatier 1987; Grant 1996; Marzotto et al. 2000; Desfor and Keil 2004, chap. 7), but, as I have held throughout this book, primarily by that segment of the business community whose economic fortunes are tied to increasing land values and local investment.

My employment of an ethical criterion to analyze the political behavior of environmental activists is not intended to draw ethical or moral judgement or condemnation. Instead, this chapter utilizes a particular framework of ethics that also serves as an analytical framework. It elucidates the role and impact of those environmental activists who choose to operate within the polity in their attempt to affect cleaner air. Moreover, the utilization of an ethical framework rooted in notions of democracy can help determine the most efficacious usage of environmentalists' limited resources.

ECOLOGICAL MODERNIZATION,
THE AUTOMOBILE, AND SYMBOLIC INCLUSION

In his analysis of interest group inclusion within the policymaking process, democratic theorist John Dryzek explains that "oppositional groupings can only be included in the state in benign fashion when the defining interest of the group can be related quite directly to a state imperative" (1996a, 479). In other words, according to Dryzek, groups that critique the status quo can only participate in the policymaking process to the extent that the groups' goals are consistent with an objective of the state. This is reflected in the behavior of the environmental groups that are active in the formulation of California's automobile emission regime.

Environmental activists involved in this process are aware of the relationship between economic growth, a greater number of automobiles (with gasoline-combustion engines), and air pollution. As a transportation analyst for the Union of Concerned Scientists (UCS) explained:

> There's definitely I think a broad agreement throughout the environmental community that the sustainable strategy for dealing with transportation, not only from an environmental perspective, but from a social perspective is better land use management. To get people out of cars, better jobs, housing balance, renewal, urban centers, density—I think all of the buzz words come to bear. (Mark 2000)

The executive director for the California Coalition for Clean Air (CCCA) argued that in terms of improving air quality in the state, "I think there are a considerable number of environmentalists that think that limiting growth is a good idea" (Carmichael 2000). A senior staff attorney for the Natural Resources Defense Council (NRDC) noted the obvious relationship between growth, increased automobile usage, and air pollution when she stated that "more cars" in an area equals "more pollution. It's a cycle" (Hathaway 2000).

Despite the relationship observed between growth, a growing number of automobiles (with gasoline-combustion engines), and air pollution, these activists also acknowledge that the idea of regulating growth and the number and usage of automobiles to address air pollution is not considered as a policy option in the policymaking process. The official from the UCS explained "that motor vehicle [air emissions] policy is thought of relatively separately from transportation [and] land use policy" in California politics (Mark 2000). The person from the CCCA stated that the government's "pollution control strategy is focused on technology" (Carmichael 2000). When asked whether regulations on economic growth within the policymaking process were actively considered in relationship to automobile emissions, air pollution, or other environmental concerns, the NRDC senior attorney unequivocally stated that growth "is currently, absolutely off the table" (Hathaway 2000).

With the realization that regulations on growth and on the number of automobiles have no realistic chance of being imposed to improve air quality, environmental activists limit their political activity to the promotion of technology to address the automobile as a source of air pollution. This is reflected in my interviews with activists from environmental groups that participate in the policy formulation process as it relates to automobile emissions in California (Mark 2000; Carmichael 2000; Hathaway 2000): the California Coalition for Clean Air, Natural Resources Defense Council, and the Union of Concerned Scientists. By promoting technology as the means to reducing air pollution from the automobile, these environmental groups and activists are simply promoting the ecological modernization of the automobile, which is wholly consist with the state imperative described above. At best, they can be viewed as the most aggressive advocates within the policymaking process of the automobile's ecological modernization. This is congruous with Dryzek's observation that when the state has imperatives oppositional groups within the policymaking process are limited to "influencing how imperatives are met, and how trade-offs between competing imperatives are made" (1996a, 480).

Therefore, in order to be effective within the policy formulation process, environmentalists, within this specific context, must drop their critiques of economic growth and the increased usage of the automobile. As explained by Dryzek, under a political process where the state has imperatives, "a high price will be paid by any [oppositional] group included [within the state] on this basis. For if state officials have no compelling reason to include the group, then presumably it must moderate its stance to fit with established state imperatives" (1996a, 480). This reality was most cogently conveyed by the official within California Air Resources Board (CARB) who is directly in charge of the state's motor vehicle emission program when he stated that environmental activists became much more effective within the policymaking process once they dropped their critiques of growth. He specifically stated:

> In the past there was some tendency [among environmentalists] to be staking out a philosophical view and perhaps being a little more strident in looking for compromise. In the last ten years or less, I think the environmental community has become much more of a court player.

He went on to laud environmentalists because currently they are "much more focused on solutions and less on philosophical issues. I mean, for example, . . . people used to argue for . . . no growth." This official considered the current approach among environmentalists within the policymaking process to be "pragmatic." He also felt that today environmentalists "seem to look for compromises and be part of the solution" (Cackette 2000).

REBELLIOUS POLITICS, THE ENVIRONMENTAL LOBBY, AND DEMOCRACY

As I noted in chapter 1, environmental activists' participation in the policy formulation process lends legitimacy to it (Edelman 1977; Saward 1992). This process, however, is undemocratic. This is because certain actors or forces block central political and economic issues from making it onto the agenda (Bachrach and Baratz 1962; Crenson 1971; Lukes 1974; Lindblom 1982; Hayward 2000). The lending of legitimacy to this process serves to undermine the potential development of a substantive and democratic politics where such issues as economic growth and the usage of the automobile can be seriously debated.

Specifically, by participating in the government's policymaking process certain environmental groups are helping to prevent the development of a broader social movement in civil society that could challenge and debate the imperative of growth. Jeffrey Isaac defines civil society as "those human networks that exist independently of . . . the political state" (1993, 356; also see Isaac 2003). Dryzek argues that civil society is a more democratic venue than the state, because it "is relatively unconstrained." He goes on to explain that within civil society:

> Discourse need not be suppressed in the interests of strategic advantage [as is the case within the state]; goals and interests need not be compromised or subordinated to the pursuit of office or access; embarrassing troublemakers need not be repressed; the indeterminacy of outcome inherent in democracy need not be subordinated to state policy. (1996a, 486)

Thus, democracy here is defined as the ability to consider and advance an indeterminate number of policy means and goals. Hence, Dryzek holds that this openness can only take place outside of the state, because the state is tied to specific objectives.

To the extent that some of the possibilities considered and advanced within civil society contest and confront the state's imperatives, Isaac avers that within civil society "rebellious" politics can take place. He holds that

> [a] rebellious politics is a politics of voluntary associations, independent of the state, that seeks to create spaces of opposition to remote, disempowering bureaucratic and corporate structures. Such a politics is often directed against the state, but it does not seek to control the state in the way that political parties do. Neither does it lobby the state to achieve specific advantages, as do interest groups. Rather, it is a politics of moral suasion, seeking . . . to affect the political world through the force of its example and through its very specific, proximate results. (1993, 357; also see Kohn 2003)

When they achieve critical mass, rebellious politics are transformed into broad-based social movements. Sidney Tarrow explains that social "movements mount challenges through disruptive direct action against elites, authorities, other groups or cultural codes" (1994, 4).

Within U.S. civil society, and disconnected from the state, exists a rebellious environmental politics that challenges economic growth and the priority this growth is given above other values, such as human health, the humane treatment of animals, and environmental sustainability. This politics is led by Earth First!, animal liberationists, and networks organized around toxins and environmental justice (Bullard 1990; Szasz 1994; Dowie 1995; Taylor 1995; Dryzek 1996a, 480; Schlosberg 1999; Wall 1999; Doherty 2002).

Dryzek asserts that "whether a group should choose the state, civil society, or both simultaneously depends on the particular configuration of movement interests and state imperatives" (1996a, 485). He goes on to aver that the most efficacious approach for the environmental community to take is a "dualistic" approach (Cohen and Arato 1992; Wainwright 1994), where part of the community operates within the state to advance the ecological modernization of capitalist society. The more confrontational portion of this community should then operate largely within civil society where they can confront the imperative of growth, and its attending environmental ill effects. Moreover, the activities of the more contentious portions of the environmental community, by placing outside pressure, can help that portion within the state to advance the goal of ecological modernization (Dryzek 1996a, 483–486). The difficulty with this dualistic approach is that it fails to take into account how incorporation within the state can serve as a means to undermine rebellious politics and social movements in civil society.

THE CONTAINING OF REBELLIOUS POLITICS

Historically, the state has not been passive in the face of rebellious politics and the emergence of social movements. Instead, it attempts to ensure that rebellious politics do not achieve critical mass, which could destabilize society or force the state to substantially alter its imperatives as a concession to confrontational social movements (Tarrow 1994). One means to contain rebellious politics is through coercion (Sexton 1991; Acher 2001).

Another means is to "buy off" those groups and individuals that could potentially be part of a rebellious politics. Progressives, socialists, and Marxists have historically viewed mainstream labor unions and welfare programs as overt attempts on the part of the state and corporations to blunt class conflict and politically subdue and pacify the working class to maintain internal order (Weinstein 1968; Piven and Cloward 1971; Domhoff 2002). Maintaining internal order is a key imperative of the state (Skocpol 1979).

Certain critical thinkers, as I explained in chapter 1, argue that the state manages the public's environmental concerns primarily through the dissemination of symbols (Edelman 1964; O'Connor 1994; Cahn 1995). Cahn

(1995) specifically avers that the federal government's post-1970 environmental regulatory policies (i.e., clean air, clean water, energy, and waste policies) can be most aptly characterized as symbolic responses to the public's growing environmental concerns, rather than as substantive efforts to regulate corporate America. He arrives at this conclusion by analyzing the content of these policies. Furthermore, Cahn juxtaposes the content of these policies with the federal government's continued encouragement of economic growth, and its continued support and subsidization of fossil fuels usage (e.g., road and highway maintenance and expansion). These are the primary factors that cause air and water pollution, as well as waste creation. Thus critics like Cahn argue that federal environmental legislation and environmental policies designed to regulate corporate America are symbolic precisely because they do not challenge the state's imperative of economic growth, nor have they sought to alter the economy's reliance on highly polluting fossil fuels, especially gasoline as an automotive fuel.

In the California context, long-term regulatory planning by state agencies can also be interpreted as a symbolic response to the public's environmental concerns. For example, the CARB promulgated a plan in 1990 that mandated that 2 percent of automobiles offered for sale in 1998 be Zero Emission Vehicles (ZEVs), 5 percent by 2001, and 10 percent by 2003 (Kamieniecki and Farrell 1991; Grant 1996). Currently, only electrically powered vehicles have zero emissions. Similarly, California in 1989 adopted the Air Quality Management Plan (AQMP) (Kraft 1993). The state's AQMP also relied heavily on the long-term development of technology to achieve improvements in air quality. Significantly, neither of these plans put forward subsidies to facilitate the development of hoped-for technologies, nor did they mandate sanctions for industrial sectors that failed to develop the necessary technologies. Commenting on the state's AQMP shortly after it was promulgated, Sheldon Kamieniecki and Michael Farrell astutely noted that "for mainly political reasons, the more difficult decisions [of the AQMP] have been postponed for a number of years, with the hope that new technologies will allow policymakers to meet federal clean-air standards with minimum disruption to . . . economic growth" (1991, 154). Notably, the targets for the manufacture and sale of ZEVs have been postponed and reduced significantly by CARB (Cone 1995 Dec. 7; 1995 Dec. 20; 1995 Dec. 22; Hakim 2003 April 25).[2] In 2002 California enacted a law mandating the reduction of greenhouse gas emissions from automobiles. The law does not go into effect, however, until 2009 (Cushman 2002).

The group mobilization incentive structure outlined by Olson (1971) offers part of the explanation as to why, even in the face of persistently poor air quality, rebellious politics in California, or throughout the United States, have not been transformed into a social movement. The symbols emanated with the enactment of regulatory legislation, and unenforced regulatory frameworks, as suggested by Edelman (1964; 1988), also contribute to the public's relative political passivity on the issue of air pollution.[3] These symbols communicate to the public that something is already being done to address the issue of air pollution and that they need not spend their time and energy attempting to overcome the collective action barriers inherent in the mobilization of large groups.

To the symbols emanated with the passage of environmental legislation, and regulatory guidelines that push difficult decisions into the future, can be added the symbolic inclusion of environmental groups within the policy-making process. In other words, the benign inclusion of environmental activists within the policymaking process that produces California's automobile emission standards becomes part of the symbols deployed against the public, and works to keep it demobilized on the issue of air pollution. The participation of environmental activists in the policymaking process communicates to the broader public that this process is democratic, because it is putatively inclusive of all relevant political perspectives (Edelman 1977; Wynne 1982; Saward 1992).[4] Environmental activists' participation in the California polity takes the form of both formal and informal access to the Governor, state legislature, and the CARB (Mark 2000).

As I have already argued, this policymaking process is not democratic, because the key issues of economic growth and automobile usage are kept off the agenda.[5] Moreover, environmental concerns are addressed within this process only to the extent that they do not conflict with those economic interests that monetarily benefit from increasing land values and the increasing sale of automobiles and gasoline (Davis 1993; Luger 2000; Olien and Olien 2000). This is especially evident as the CARB has backed away from policies designed to force the development, production, and distribution of alternative fuel automobiles.

THE ENVIRONMENTAL LOBBY AND CALIFORNIA'S ECOLOGICAL MODERNIZATION OF THE AUTOMOBILE

Why would environmental activists want to lend legitimacy through their participation to an undemocratic policy formulation process? A more intuitive question might be why would environmental activists want to participate in a process that keeps central issues from being effectively discussed? One answer to this question is that the ecological modernization of the automobile would not be going forward were it not for the participation of these activists. All of my environmental activist interviewees, including the contracted lobbyists for the Sierra Club, did feel that their specific role in the policymaking process was to put political pressure on government actors and prompt the strengthening of the state's automobile emission standards. Implicit in their thinking is that were it not for their participation within the policymaking process CARB officials would not have the same incentives to reduce emissions from automobiles.

One of my respondents acknowledged that the CARB and the automobile industry have a type of "symbiotic" relationship, with the industry being the primary source of CARB's technical knowledge on emission control technology and alternative-fuel automobiles, and in return the CARB only sets emission standards that the industry can comply with (White 2000; also Cackette 2000 and Scheible 2000). This same person also explained that the environmental lobbying community attempts to keep these actors "honest." In other words, this lobbying community works to apply "pressure" and serve as a "check" to a potentially cozy relationship between the CARB and the automobile industry (White 2000). Moreover, to the extent that the California environmental lobbyists advocate policies that would force the development, mass production, and distribution of alternative-fuel automobiles, my respondents from this community felt that they were moving the CARB and the automobile industry into a technological direction that these actors would not pursue on their own.

Whether the inclusion of environmental activists into the policymaking process actually results in the increasing ecological modernization of the automobile is unknown, however. As I pointed out in chapter 5, policies advancing the ecological modernization of the automobile preceded the inclusion of environmental activists into the policymaking process. The first California law requiring the installation of emission control technology in automobiles was enacted in 1960. Environmentalists were not incorporated into the policymaking process on this issue in California at least until the 1970s (Roberts 1969, chap 3; Krier and Ursin 1977, chap. 14; Fawcett 1990, 91–93; Dewey 2000, chap. 5).

As I also pointed out in the chapter 5, early ecological modernization efforts in California were led by economic interests that monetarily benefitted from local economic and population growth and increasing land values. It was the *Los Angeles Times* and its owner, Norman Chandler, that led the original campaign in the 1940s to ecologically modernize business and industry in the city of Los Angeles. This campaign resulted in the creation of the state's first pollution control agency, the Los Angeles Air Pollution Control District.

Additionally, it was an economic elite-led policy-planning organization, the Air Pollution Foundation, that in the 1950s politically established the automobile as a major source of smog in Los Angeles, and advocated the creation and application of pollution control technology to it. As noted in chapter 5, the Foundation board of trustees was composed almost entirely of representatives from business and industry. Many of these individuals represented firms that directly benefitted from economic growth in the Los Angeles basin, as well as firms involved in the production and sale of automobiles and gasoline. Moreover, the Foundation's list of contributors demonstrates the broad support that it enjoyed throughout the corporate community. Throughout its seven-year existence (1954–1961) the Foundation had more than 200 donors, almost all of which were from corporate America.

In order to formulate policy proposals and political strategies that relate to the environment, members of the contemporary California business community have organized and financed the California Council for Environmental & Economic Balance (CCEEB).[6] This organization was established in 1973. One-third of its board of directors is drawn from business and industry. Some of the firms represented on CCEEB's board are economically dependent on growth in the state. They are the Irvine Company (real estate and land development), Pacific Telesis (regional telephone service provider), Southern California Edison (utility firm), Bank of America, and Pacific Gas & Electric Company. Other firms represented on CCEEB's board are directly affected by the state's environmental regulations. These firms include Texaco, Chevron, and the Union Pacific Railroad.[7] The other two-thirds of its board is composed of labor union representatives and private citizens.[8] The CCEEB's finances are entirely provided by its corporate members (Weisser 2000). Moreover, CCEEB disseminates its policy ideas throughout the California business community through "presentations" to such organizations as the Los Angeles Chamber of Commerce, the Santa Clara Manufacturers Group, and the Orange County Industrial League (CCEEB 2000b).

The CCEEB describes its work in terms that are consistent with the concept of ecological modernization. Its president stated in our interview that a "'healthy' environment leads to a 'good' economy." He claimed that all of CCEEB's members adhere to this belief. Moreover, he explained that, for example, a clean environment plays a specific role in maintaining and attracting high-technology industries to California (Weisser 2000). According to its mission statement, the CCEEB "is a coalition of California business, labor, and public leaders who work together to advance collaborative strategies for a sound economy and a healthy environment" (2000b). In its promotional literature, the CCEEB describes itself as a "powerful coalition that can move the California economy forward in an environmentally responsible manner." Additionally, it claims to "understand the importance of the environment to the California business climate" (CCEEB 2000b). Finally, in terms wholly congruous with the notion of efficiency at the core of the discourse of ecological modernization (Weale 1992, 75–79; Hajer 1995, 31–36), CCEEB holds that its "work translates into . . . job creation, efficient use of tax dollars, reduced compliance costs, consolidated reporting formats, elimination of multiple agency oversight, more responsive public agencies and increased certainty for conservation and development" (CCEEB 2000a).

CCEEB's commitment to environmental protection within the context of unmitigated economic growth is reflected in its political activity. It supported the 1988 California Clean Air Act. The President of CCEEB proudly averred that CCEEB was "instrumental" in the passage of this legislation (Weisser 2000). The Act mandated a 55 percent reduction in automobile emissions, a 15 percent reduction in nitrogen oxides, and "lowering the levels of suspended particulates, carbon monoxide, and toxins as is technologically feasible" all by the year 2001 (Lamare 1993, 239). Additionally, the President of CCEEB explained that its members are "leaders on mobile source emission [reduction] efforts in California" (Weisser 2000). The CCEEB also touts that it recently "co-sponsored the most comprehensive package of legislative proposals for a statewide growth strategy ever advanced; the package provided a structure for more certainty for both development and conservation" (CCEEB 2000b).

THE BUSINESS COMMUNITY, URBAN GROWTH, AND ECOLOGICAL MODERNIZATION

Therefore, the public policies that ecologically modernize the California economy and the internal-combustion engine are consistent with the political activity and economic interests of that segment of the business community whose economic success depends on economic growth within the state. In light of the historic political activity of many of the members of California's growth coalition, their objective economic interests, and the acute air pollution inundating Los Angeles beginning in the 1940s (discussed in the last chapter), air pollution abatement regulations can be viewed as part of a legal infrastructure that helps attract capital to the state, and ultimately facilitates and promotes local growth, much like an education and transportation infrastructure does. While these factors may drive up the costs of production, they provide for an educated workforce, a transportation network, and a more salutary environment for workers and production.

In the contemporary period, among those interest groups involved in the formulation and implementation of California's automobile emissions standards, it is the business-led and financed CCEEB that most openly and aggressively embraces the discourse of ecological modernization. Its championing of the ecological modernization of business and industry as well as the internal combustion engine is reflected in its political activity.[9]

With leading members of the corporate community providing political support and energy to those public policies that ecologically modernization the internal-combustion engine, the political contribution of those environmental activists incorporated into the policymaking process becomes unclear. In other words, with the business community promoting the ecological modernization of the gasoline-burning internal-combustion engine, it becomes difficult to determine to what extent environmental activists are advancing the ecological modernization of the automobile through their participation in the policymaking process. Hence, those public policies designed to forward the ecological modernization of the internal-combustion engine could be the result of business political actions, and not those of environmental lobbyists. Furthermore, determining the actual influence of environmental lobbying efforts is complicated by the fact that, as Dowie (1995, 48) points out, environmental groups, for purposes of fund raising, are apt to take full credit for perceived legislative or regulatory victories even when they do not deserve it. Additionally, with the CARB backing away from its policies designed to force the mass production and distribution of alternative fuel automobiles, the notion that incorporated environmental groups exercise significant influence over this agency appears as doubtful. This agency has done so in face of strong opposition from the environmental lobbying community (Carmichael 2000; Hathaway 2000).

Moreover, given the political activity of the CCEEB and the local economic interests adversely affected by air pollution, what does become apparent is that the ecological modernization of the gasoline-burning internal-combustion engine will go forward with or without the incorporation of environmental activists. Albeit, this modernization *might* not proceed at the same pace.

CONCLUSION

In light of the factors outlined here, the participation of environmental activists in the policymaking process takes on ethical dimensions. What we can see is that these activists serve to enhance public support for a policy formulation process that primarily abets the economic needs of the business community, while also dampening those political forces that would compel the treatment of questions and issues that are central to a salutary and sustainable environment. Moreover, it is uncertain what environmental activists gain, in terms of environmental protection, for their participation in the policymaking process.

The environmental community as a whole, however, has affected politics. Its gains are most readily evident in the realm of public opinion. Environmental ethicist Lester Milbrath (1995) explains that "public opinion polls show that a majority, usually a high majority, of people in most countries are aware of environmental problems and very concerned about getting them solved to ensure a decent future." (102). No doubt, the environmental community deserves at least partial credit for the awareness among the world's citizenry of environmental problems (Desai 2002; Smith 2002; Guber 2003).

This success, along with the dubious nature of their participation within the policy formulation process, would suggest that the most efficacious deployment of the environmental lobbying community's resources would directly involve the public and specifically civil society. As Milbrath explains, environmental "awareness and concern does not necessarily mean that people well understand" the causes and potential solutions to society's environmental ills (1995, 102; also see Bednar 2003). Hence, instead of maintaining a somewhat hostile and contentious attitude toward those confrontational environmental groups and networks that operate outside of the polity (Dowie 1995; Doherty 2002), the environmental lobbying community should exit the polity and join with their rebellious brethren in civil society (Norton 1991). In this way, those resources currently deployed lobbying officials within government could be more fruitfully directed at educating the public.

Such an education effort would involve informing the public how government agencies, both federal and state, utilize a narrow or "weak" conception of ecological modernization to address air pollution (Christoff 1996; Dryzek 1997, chap. 8; Neumayer 2003). A narrow approach to ecological modernization relies heavily on technological solutions to address pollution. A more expansive or "strong" conception of ecological modernization would involve ecologically sensitive land management. This type of land management would entail the intensive usage of land (as opposed to sprawl [Purdum 2000]), drawing residential and work areas closer together, and making mass transit the primary means of rapid transportation in urban areas. Ecologically sensitive land management would move residents away from their dependence on the automobile (and the gasoline-combustion engine) and toward more ecologically benign forms of transportation, such as walking, bicycling, and mass transit (Newman and Kenworthy 1999; Pinderhughes 2004).

Additionally, such an education campaign could serve to expand environmental rebellious politics into a social movement that could potentially force policymakers to abandon the narrow version of ecological modernization and instead employ the more expansive version of this concept. While it can be debated how practical, realistic, or desirable it is for the environmental lobbying community to redirect its resources away from the polity and toward civil society, one thing is clear. Viewing it through the lens of democracy, such a redeployment of resources is ethical.

CONCLUSION

Political Power and
Global Warming

In chapter 1, I outlined how U.S. clean air policies are market enhancing. Statistical studies have demonstrated a positive relationship between state-level clean air policies and state-level economic activity. These policies are market enhancing precisely because they help protect local economies from the negative impact of air pollution. Such policies do so by providing cleaner air than otherwise, and thus help to attract investment and people to an area. Clean air policies play this market-enhancing role because policymakers rely on technology in formulating them. Pollution control technologies help to abate localized air pollution, but do not directly affect the rates of local economic growth. In addition, such an approach to air pollution abatement is acceptable for industry because it develops and deploys pollution control technologies.

The key question for political scientists is what factors have prompted the reliance on technology to reduce air pollution, and what factors shape clean air policies? In chapter 2, I put forward two competing policymaking models to analyze the political factors that have historically determined the response in the United States to localized air pollution. These models are entitled state autonomy/issue networks and economic elite theory. The first emphasizes the ability of officials within the state to formulate and implement public policies independently of societal groups, including business interests. Additionally, these officials will often draw policy ideas from different groups, experts, and perspectives. In this way, environmental groups and their views are often incorporated into the policymaking process.

In contrast to the state autonomy/issue networks model, the economic elite model stresses the role of economic elites in policy formulation. These elites exercise dominant influence over the state because they possess ample wealth and income, which can be converted into a number of useful and advantageous political tools: campaign finance, publicity, access, etc. Moreover, economic elites operate through policy-planning organizations, which supply them with information, analysis, and the means to build a consensus on proposed policies among elites in general. Through such organizations, economic elites can determine and agree upon those policies that are going to enhance their market positions and improve the overall operation of the economy.

In turning to the empirical record, it becomes evident that the economic elite model offers deeper insight into the development of air pollution abatement policies than the state autonomy/issue networks one. As outlined in chapter 3, local businesspeople, and local chambers of commerce, have historically perceived air pollution as an economic negative. During the late nineteenth and early twentieth centuries, prominent business people in Chicago took the political lead in the effort to abate air pollution. In seeking to do so, they operated through the Chicago Association of Commerce (CAC) and the Society for the Prevention of Smoke, which were comprised of prominent business people—many of whom were heavily vested in the local economy. Both groups sought to use technology to improve the city's air quality. Apart from local chambers of commerce and certain economic elite-led organizations, only groups largely made up of and led by upper-class women undertook organized political action to address urban air pollution in this era. Like the CAC and the Society for the Prevention of Smoke, they centered their anti-smoke agenda on pollution control devices.

Through the efforts of the CAC and the Society we can understand why locally oriented economic elites' concerns during the late nineteenth and early twentieth centuries about air pollution did not result in cleaner urban air during this period. Railroad firms were generally opposed to any effort to force the electrification of their lines. Technological controls on coal-generated smoke from stationary sources were labor intensive and often ineffective. Hence, the implementation of technology to control air pollution in either instance would have directly and negatively affected economic growth in Chicago and other localities. Without technology as a viable means to reduce air pollution, those U.S. industrial cities that were heavy users of soft coal remained inundated with air pollution into the middle of the twentieth century. The response of public officials in cities with acute air pollution was to either ignore the problem or assuage the public through rhetoric, unimplemented legislation, token enforcement actions, and/or ineffective regulatory agencies.

In the mid-twentieth century, oil and natural gas became economically viable fuels for the country as a whole. As a result, industrial cities that heretofore burned large amounts of soft coal now experienced significant improvements in air quality. During this period, however, a new threat to air quality became manifest: the automobile. As explained in chapter 4, the automobile became the dominant form of transportation in urban America in large part because real estate interests (i.e., large land holders and developers) viewed it as a relatively low cost way to bring utility to land throughout metropolitan areas. While the automobile brought definite economic benefits to key members of local growth coalitions throughout the United States, the mass usage of the automobile also brought acute air pollution for many localities.

It was in Los Angeles, among major cites, where land developers on a general basis first integrated the automobile into their development projects. As a result, by the 1920s, Los Angeles residents were leading the nation in the purchase and use of the automobile. This, combined with its rather unique topography and meteorology, led the Los Angeles basin to be the first area in the United States to experience severe air pollution from the automobile.

In chapter 5, I pointed out that it was those economic interests that benefitted from economic growth in Los Angeles that took the most direct steps to comprehensively deal with the air pollution plaguing the region beginning in the 1940s. A key part of this work took place through the *Los Angeles Times* Citizens Smog Advisory Committee, the Air Pollution Foundation, and the Los Angeles Chamber of Commerce. Feminine members of the Los Angeles upper class also played a prominent role in air pollution abatement during this period. They did so through the organization known as SOS (Stamp Out Smog). Like Chicago at the turn of the century, economic elites in Los Angeles sought to manage the air pollution in their area through the deployment of technology. This included air pollution from the automobile. Unlike the Chicago efforts at air pollution abatement, however—which failed to produce substantive results—some success was achieved in the Los Angeles effort to reduce air pollution. This is especially because automotive pollution control technology was both effective in reducing pollution *and* relatively inexpensive. Moreover, this expense could in significant part be passed on to the consumer. Additionally, with technological solutions seemingly viable approaches to the air pollution situation in Los Angeles, the apparent effectiveness of those regulatory policies forwarding the development of technologies to address air pollution could be publicized.

Outside of the economic elite and state autonomy/issue networks models, other approaches have been developed to explain why U.S. clean air policies rely upon technology to reduce air pollution. These different approaches can be identified as "policy learning" (Sabatier 1987; 1999), neo-Marxist (Barrow 1993, chap. 2; Aronowitz and Bratsis 2002), and ecological modernization discourse (Weale 1992; Litfin 1994; Hajer 1995; Bernstein 2001).

To one degree or another, they all point to the political activity of modern environmental groups to explain the content of air pollution abatement policies. In the case of the policy learning approach, public policies mandating pollution control technologies to manage air pollution are the result of the interaction and competition between industry and its supporters, on the one hand, and environmental groups and their allies, on the other hand. Through this interaction and competition, both groupings come to accept "clean" technologies as an adequate compromise. In the case of the ecological modernization approach, its proponents hold that the political dynamic posited in the policy learning approach has resulted in the generally held notion or discourse that technology can and should be deployed to make economic activity environmentally benign.

Finally, advocates of the neo-Marxist view of the state posit that officials within the state seek to attain economic growth and legitimacy. Arguing within this paradigm, Dryzek (1996a; Dryzek et al. 2003) has forwarded the idea that environmental groups have successfully modified the concept of legitimacy in the United States. Now, to attain legitimacy, the state must provide some measure of environmental protection. In order to reconcile the somewhat contradictory objectives of promoting economic growth and environmental protection, officials within the state have promoted technological controls on air pollution. In this way, both of the state's objectives are pursued.

From the history presented in this book, it is apparent that the idea to deploy technology to abate air pollution was on the political agenda long before the advent of the modern environmental movement—as early as the late nineteenth century. Moreover, policymakers successfully utilized technology to reduce airborne emissions significantly before contemporary environmental groups were seeking to directly influence clean air policies. This is most apparent in the case of California.

In addition to discounting the policy learning, neo-Marxist, and ecological modernization approaches as viable explanations to account for the content of U.S. clean air policies, the fact that the technology-focused air pollution abatement agenda was formed and forwarded prior to the introduction of modern environmental groups sheds significant insight into the political impact of these groups. It indicates that these groups are incorporated into the policymaking process on the basis of an agenda that was set before their incorporation, and that environmental groups have not been able to alter this agenda. As I have been arguing, this agenda is set by local growth coalitions, industry, and the energy sector.

While a technological approach to air pollution control has improved localized air quality for many areas, several cities in the United States continue to have unhealthful air, and in some air quality is deteriorating (e.g., Houston [Dawson 1999; "Smog City" 1999; Cherni 2002]). Even more alarming, the use of technology to improve local air quality has not prevented the United States from being the leading absolute and per capita emitter of the key global warming gas: carbon dioxide. With less than 5 percent of the globe's population, the U.S. economy emits about 25 percent of the world's total emission of this gas (Warrick 1997; Revkin 2001 June 12; Uzama 2003).

With U.S. air pollution politics and policies configured predominantly to be market enhancing, it is not surprising that under the Clinton administration no substantive effort was made to comply with the Kyoto Protocol (Revkin 2000; Steinberg 2002, 273–274; Brown 2002), and the current Bush administration abandoned any effort at its implementation (Lisowski 2002). If fully implemented, the Kyoto Protocol would have required the United States to lower its carbon dioxide emissions by roughly 20 to 30 percent by 2012 (Victor 2001; Roberts 2004, 138). The effective implementation of this accord would have required a significant reform of the United States's land, energy, and/or transportation use (Newell 2000; Leggett 2001; Pinderhughes 2004). These reforms would impose significant costs on these sectors, and would not have any immediate positive economic effect (Reitan 1998; U.S. Congress 1998; Nordhaus and Boyer 2000).

Therefore, a technology-focused agenda dominates both state and federal government, and this agenda has failed to provide healthful air for millions (Kamieniecki et al. 1999; Davis 2002; Lee 2004; Herbert 2004) or prevent the current global warming trend. Given these realities, those environmental groups and activists interested in protecting the environment, and decisively addressing urban air quality issues, should withdraw from the polity and seek to mobilize the public on both local air quality and global warming issues. This would be beneficial in two regards. First, by withdrawing from the policymaking process such groups and activists would not aid in conferring legitimacy on this process. Second, by focusing their total attention and resources on the broader public, environmental groups could contribute to sparking a social movement on these issues. Additionally, as I argued in chapter 6, given the policy work and political activity of the California Council for Environmental & Economic Balance, the technology-focused clean air agenda will go forward even without the participation of environmental groups in the policymaking process.

It appears that only through a confrontational social movement will the dominant position of local growth coalitions, industry, and the energy sector over land use, energy, and transportation policies be challenged. Only through such a challenge, can democratic and decisive policies be instituted to address the health and environmental perils created by airborne emissions.

Notes

CHAPTER ONE
LOCAL GROWTH COALITIONS,
ENVIRONMENTAL GROUPS, AND AIR POLLUTION

1. The U.S. economy is the source of about 25 percent of the world's carbon dioxide emissions (Revkin 2001 June 12; Uzama 2003).

2. Barrow (1993) explains that "corporations emerged as the dominant economic institutions in capitalist societies by the end of the nineteenth century." He goes on to note that as early as the late 1920s, "the bulk of U.S. economic activity, whether measured in terms of assets, profits, employment, investment, market shares, or research and development, was concentrated in the fifty largest financial institutions and five hundred largest nonfinancial corporations" (17).

3. Despite this reduction of automotive emissions, automobile emissions continue to pose a serious health risk because there are more automobiles on the road and people are driving longer distances than in the 1960s (U.S. Congress 1990, 227 and 231; Kenworthy and Laube 1999).

4. Despite this provision, and the "no third vehicle" clause, California maintains an automotive alternative fuel plan. It is not being legally challenged (Hakim 2003 August 12).

5. The ability of the EPA to monitor compliance with the reformulated gasoline standards is hampered by the fact that the oil industry was successful in having the federal government utilize an averaging scheme to measure compliance, as opposed to a gallon-by-gallon approach (Weber 1998, chap. 5; Gonzalez 1999).

6. The Clean Air Act of 1990 did contain one clearly nontechnological approach to air pollution abatement. The Act has a provision calling for an increased emphasis on car-pooling to improve urban air quality. This provision did not amount to much, however, since the Act does not provide for sanctions in cases of noncompliance (Lee 1996). It was not initiated by environmental groups, but by officials from California (Marzotto et al. 2000, chap. 4).

CHAPTER TWO
POLITICAL ECONOMY AND THE POLICYMAKING PROCESS

1. The concept of political economy that informs my argument is one that views economic systems as political systems (Block 1990; Roy 1997). This is because economic systems historically privilege certain values and interests, often at the expense of other values and interests. Capitalism—of particular relevance to this study—prioritizes profit and the interests of the capitalist class (i.e., the economic elite), often at the expense of workers and the environment (Gorz 1994; Perelman 2003). In light of this, a central question of this study is how have the environmental policies in question affected capitalism both as an economic and political system? Specifically, to what extent have environmental policies interjected the values of environmental protection into the operation of the economy and shielded the ecosystem from damage created by economic activity? Or, have these policies largely forwarded the values, economic and political interests, and preferences of economic elites and producer groups at the expense of comprehensive environmental protection and the global ecosystem?

2. Almond (1988) contends that the claim of originality made by state autonomy theorists is unwarranted, since autonomous state officials have been an explicit aspect of pluralist theory from its inception.

3. I offer a full description this model elsewhere (2001a, 10–13), so here I will only provide its central features.

4. Among the environmental interest groups that Shaiko examines in detail, it is ED that most overtly takes the position that its membership is primarily, if not exclusively, an economic resource for its leadership. While the NWF does not openly embrace ED's view of leadership-member relations, it nonetheless limits its political communications to its members for fear of alienating existing and potential dues paying members. The NWF by far has the largest membership among environmental groups (Shaiko 1999, 41). Among the groups that Shaiko studied, EA was the most active in seeking to mobilize its membership to affect political change. Significantly, EA historically maintained a relatively small membership base, and in 1996 went defunct due to insufficient financial resources. The SC and TWS make more concerted efforts to communicate to its membership on political issues than either ED or the NWF. Additionally, the SC maintains an institutional mechanism to allow its members to communicate to the group's leaders on issues of public policy. Shaiko nonetheless concludes that in the contemporary period the leadership of these groups prioritize organizational maintenance over political efficacy (Gonzalez 2000).

5. I provide a full review of the plural elite literature elsewhere (Gonzalez 2001a, 2–10).

6. Please see note 2 of chapter 1.

7. The economic elite-led policy-planning network has two groupings—one that is characterized as "moderate" or "corporate liberal" and the other as "conservative." While these two groups will frequently compromise on issues, they sometimes cannot. When they cannot find common ground, their struggles will usually spill over into

government where each will utilize its political strength to try and get its way (Weinstein 1968, chap. 1; Eakins 1969; 1972; Domhoff 1978a, chap. 3; 1990, 38–39; Barrow 1993, chap. 1).

CHAPTER THREE
THE POLITICS OF AIR POLLUTION DURING THE
LATE NINETEENTH AND EARLY TWENTIETH
CENTURIES: THE FAILURE OF TECHNOLOGY

1. Louise Walker, in her 1941 dissertation on the Chicago Association of Commerce, notes that the Association "was formed in 1904, with ninety-three members" and "that its peak membership was about seven thousand [in 1924], and that its present size is about four thousand." She adds that the CAC's "members are drawn from the employing class and represent conservative business interests" (1). Walker also explains that

> working under the assumption that any activity which improves business conditions in Chicago is a legitimate concern of an Association of Commerce, the organization [the CAC] has given such matters as improved paving, endorsement of charities, crime control, educational surveys, and clean-up campaigns its continuous interest. It has caused special studies to be made of the stockyard, smoke abatement, housing, and the accounting methods of the Sanitary Commission. (2–3)

2. Stradling (1999) estimates that the committee's report only dedicated 17 pages to discussing "smoke's negative effects on human health, vegetation, and property within the city, but it did not attempt to determine the costs of these effects to Chicagoans" (128–129).

3. Dewey (2000) and Stradling (1999) explain that urban air pollution throughout the late nineteenth and early twentieth centuries was widely conceived as both an aesthetic and health concern (also see Platt 1995).

4. Here I am arguing that we need to analyze environmental legislation on two different planes. The first plane is the one in which such legislation guides, influences, and/or restricts policymakers and the courts. Hence, analysis via this plane examines how legislation effects or fails to effect the development of public policy. The second plane focuses on how the passage of environmental legislation affects public opinion. Therefore, even if legislation has no impact on policy it can still affect public opinion. Moreover, a particular piece of environmental legislation can have substantively divergent impacts on policy and on public opinion. The passage of a certain law, for example, may communicate to the public that a special interest is going to be effectively regulated, whereas the policy resulting from such legislation may actually strengthen the economic and/or political position of said special interest (Edelman 1964; 1988; McConnell, 1966; Stigler 1971; Kolko, 1977; Lowi, 1979; Gonzalez 2001a). At this point in my argument, I am focusing on the second plane of legislation.

5. Another factor that contributed to the inability of the working class to politically mobilize against air pollution during this period was the public argumentation often made by the opponents of government mandated smoke abatement. Namely, that such government policies undermined job creation and created an incentive for capital flight from an area (Stradling 1999; Dewey 2000).

6. An example of token, but high profile, enforcement occurred in Chicago when "over . . . several years the [Pennsylvania] railroad accumulated $10 fines in seven cases." Stradling and Tarr (1999) explain that while these "fines were small, the publicity accompanying them was strongly negative" (688).

7. Both Grinder (1980) and Stradling (1999) offer engineers as an autonomous interest group in the air pollution debate. At times such professionals did advocate for government regulation of air pollution emissions. As Dewey (2000, 8) argues, however, given their dependence on industrial firms for employment, it is difficult to see such professionals as politically autonomous. He points out that such professionals regularly offered the technical rationale for industry opposition to government mandated smoke controls. Moreover, David Noble (1977) demonstrates that the engineering profession, virtually from its inception, was shaped politically by corporate firms and economic elites. Reflective of this, to the extent that engineers and their professional organizations argued for the reduction of smoke emissions, they exclusively proposed technological solutions (Grinder 1980; Stradling 1999; Dewey 2000), as opposed to questioning the practice of locating factories and investments in already highly polluted areas.

CHAPTER FOUR
REAL ESTATE AND THE RISE OF THE AUTOMOBILE

1. In addition to real estate interests advocacy for the suburbanization of urban populations, progressive activists and government officials argued that urban congestion was a central cause of crime, poverty, and maladies. As a result, these actors encouraged trolley firms to expand their lines into sparsely populated zones in order to ease congestion. Moreover, urban middle-class elements often agitated for the introduction of trolley cars into undeveloped land so they could escape the overcrowding and pollution that characterized city centers.

2. Leading the political fight against fare increases were real estate interests and the owners of retail outlets. Real estate concerns saw inexpensive transportation as key in the sale of outlying subdivisions. Retail firms viewed inexpensive trolley transportation as central to bringing in customers from affluent suburbs (McShane 1974, 31–32; Crump 1988, 200–201).

3. In London, for example, the city government recently instituted a heavy tax for those drivers who take their automobiles into the central city district during business hours (Kennedy 2003).

4. The one business group in Chicago, and other urban areas, that looked negatively upon urban sprawl were downtown interests, who saw sprawl as undermining their position as the center of commerce for the city (Fogelson 2001).

5. The one major exception to this trend occurred in Toronto, where the trolley continued to thrive into the 1960s (Davis 1979).

6. Foster (1981) explains that "Chicago spent the staggering sum of $340 million over a thirty-year period on street widening alone. . . . That was more than twice the estimated cost of a comprehensive subway system at 1923 prices" (93).

CHAPTER FIVE
THE ESTABLISHMENT OF
AUTOMOBILE EMISSION STANDARDS

1. California's clean air policies originally grew out of efforts to improve air quality in the Los Angeles basin during the 1940s and 1950s. Thus, the environmental conditions and politics that I describe here relate mostly to the Los Angeles metropolitan area, but, for legal and practical reasons, these factors resulted largely in statewide laws, policies, and regulatory bodies (e.g., the California Air Resources Board).

2. The Los Angeles basin was chosen as the endpoint of the Southern Pacific's main southern California line because the principal owners of the railroad wanted to protect their real estate investments in San Francisco. It was their belief that Los Angeles, with its limited port facilities, would make for a poor economic competitor against San Francisco (Fogelson 1967, chap. 3; Jaher 1982, chap. 6).

3. Citizen groups are defined as organizations whose membership is open to all, regardless of place of employment, occupation, ethnicity, or religion (Walker 1991, 33–35).

4. Given that most of the firms who came out in favor of Proposition 5 were concentrated in southern California (Whitt 1982, 131), it would seem that they were hoping that the money made available through Proposition 5 could be used to build a BART-type transit system in Los Angeles (Whitt 1982, chap. 3).

CHAPTER SIX
DEMOCRATIC ETHICS, ENVIRONMENTAL GROUPS,
AND SYMBOLIC INCLUSION

1. Reflective of the policy community approach, a substantial portion of the data for this paper is composed of interviews from individuals drawn from environmental and business organizations, as well as the government officials directly involved in California's automobile emission policy community. I determined the relative importance of these organizations and individuals through a search of the *Los Angeles Times*'s computer data base for the years 1990 through 1999 using the key words VEHICLE EMISSIONS. The other method I utilized to determine the members of the state's automobile emission policy community was telephone and in-depth personal interviews. Here I would ask respondents to identify those organizations and individuals regularly involved in the formulation of the state's automobile emission standards. In total, eight unstructured interviews were conducted. They averaged approximately an hour and a half in length.

2. The current zero automobile emission plan put forward by CARB mandates that automotive firms in aggregate make available for sale in California 250 hydrogen fuel cell powered vehicles between the 2005 and 2008 model years; 2,500 from 2009 to 2011; and 25,000 from 2012 to 2014. Firms can comply with these aggregate targets with battery powered vehicles (Hakim 2003 April 25).

3. Please see note 3 from chapter 3.

4. In addition to the overtly political factors outlined here, psychological, cultural, and ideological barriers exist that prevent a more robust and confrontational politics arising from civil society to challenge the state's commitment to environmentally deleterious growth (Milbrath 1989; 1995; 1996; Cahn 1995; Bednar 2003).

5. As I have already explained, the definition of democracy I am utilizing here emphasizes the ability to consider an indeterminate number of policy goals and means. Another definition of democracy could focus on the procedure through which officials are chosen. In this approach to democracy, as long as central policymakers are democratically elected, then the policy outcomes of their decisions are legitimate and inherently democratic. This legitimacy would extend to those policymakers appointed by the democratically elected officials. Thus, as democratically elected and legally appointed officials, they are justified in excluding certain philosophical and policy perspectives from the policymaking process. This is because these officials can legitimately claim that they speak for the majority of citizens.

The proponents of the former version of democracy would retort that the electoral process in most cases in-and-of-itself does not necessitate or justify elected officials from eliminating policy options from the policymaking process. Instead many of these theorists hold that U.S. society's reliance on the market to produce and distribute goods and services results in the effective elimination of various policy options from the policymaking process (Lindblom 1982; Barrow 1993, chap. 2; Aronowitz and Bratsis 2002; Bednar 2003). Other thinkers hold that certain policy options are not considered in the policymaking process because particular political and economic interests are able to block their consideration (Bachrach and Baratz 1962; Barrow 1993, chap. 1; Hayward 2000).

6. The policy work of the Council is conducted through its "projects." According to its promotional literature, the center of CCEEB's activity is its "project committees where members design strategies which are implemented by CCEEB staff and expert consultants." This literature goes on to claim that "the result is public policy which adds value to our members rather than adding costs" (CCEEB 2000a). Project members meet once a month and projects are composed largely of various business and industry representatives and focus on specific policy areas (Lucas 2000; Weisser 2000). The work of the individual projects is augmented with "CCEEB sponsored conferences, seminars and retreats, [where] California legislators, regulators and administrative officials hear" what CCEEB members "need" (CCEEB 2000a).

7. The CCEEB project that treats automobile emissions is known as the Transportation, Emissions, and Mobility (TEAM) Project. The project is mostly composed of representatives from oil firms, utilities, and real estate concerns. While the automobile industry is not officially part of CCEEB or the TEAM project, as the project leader explained, TEAM does have regular contact with individuals from this industry (Lucas 2000). Overall, the automobile industry, which does not have any major production facilities in the state, maintains a low political profile in California. This industry largely limits its political work on the issue of automobile emissions to its relationship with the CARB. Otherwise, the automobile industry allows the oil industry, which does have several refineries in California, and local automobile dealerships to generally take the political lead on transportation issues in the state (Carmichael 2000; Hathaway 2000; White 2000).

8. Several of the union representatives on CCEEB's board are from the building trades, which have historically been strong proponents of local growth (Logan and Molotch 1987, 81–82). Moreover, many of the private citizens on its board were formerly corporate executives. One, for example, was formerly a Vice President of Bank of America (Weisser 2000). None of CCEEB's board members are leaders from the environmental community.

9. CCEEB is opposed to any public policy that mandates the usage of any specific fuel (CCEEB 1990; Weisser 2000).

Bibliography

Acher, Robin. 2001. "Does Repression Help to Create Labor Parties? The Effect of Police and Military Intervention on Unions in the United States and Australia." *Studies in American Political Development* 15 (fall): 189–219.

Ackerman, Bruce, and William T. Hassler. 1981. *Clean Coal/Dirty Air*. New Haven: Yale University Press.

Adler, Jonathan H. 1992. "Clean Fuels, Dirty Air." In *Environmental Politics: Public Costs, Private Rewards*. Edited by Michael S. Grave and Fred L. Smith Jr. New York: Praeger.

Ainsworth, E. 1946 Nov. 9. "St. Louis Has Key to Smog." *Los Angeles Times*, pt. 2, p. 1.

———. 1947 May 18. "The Showdown on Smog." *Los Angeles Times*, pt. 2, p. 4.

Air Pollution Foundation. 1961. *Final Report*. San Marino, Calif.: Air Pollution Foundation.

Alm, Leslie. 2000. *Crossing Borders, Crossing Boundaries: The Role of Scientists in the U.S. Acid Rain Debate*. Westport, Conn.: Praeger.

Almond, Gabriel A. 1988. "The Return to the State." *American Political Science Review* 82, no. 3: 853–874.

American Lung Association. 2004. *American Lung Association: State of the Air 2004*. New York: American Lung Association.

Anderson, Terry L., and Donald R. Leal. 2001. *Free Market Environmentalism*. New York: Palgrave.

Andrews, Richard N. L. 1999. *Managing the Environment, Managing Ourselves: A History of American Environmental Policy*. New Haven: Yale University Press.

Aronowitz, Stanley, and Peter Bratsis, eds. 2002. *Paradigm Lost: State Theory Reconsidered*. Minneapolis: University of Minnesota Press.

Bachrach, Peter, and Morton Baratz. 1962. "Two Faces of Power." *American Political Science Review* 56, no. 4: 947–952.

Barrett, Paul. 1983. *The Automobile and Urban Transit*. Philadelphia: Temple University Press.

Barringer, Mark. 2002. *Selling Yellowstone: Capitalism and the Construction of Nature.* Lawrence: University of Kansas Press.

Barrow, Clyde W. 1990. *Universities and the Capitalist State: Corporate Liberalism and the Reconstruction of American Higher Education, 1894–1928.* Madison: University of Wisconsin Press.

———. 1992. "Corporate Liberalism, Finance Hegemony, and Central State Intervention in the Reconstruction of American Higher Education." *Studies in American Political Development* 6 (fall): 420–444.

———. 1993. *Critical Theories of the State.* Madison: University of Wisconsin Press.

———. 1998. "State Theory and the Dependency Principle: An Institutionalist Critique of the Business Climate Concept." *Journal of Economic Issues* 32, no. 1: 107–144.

Bartik, Timothy J. 1991. *Who Benefits From State and Local Economic Development Policies?* Kalamazoo, Mich.: W. E. Upjohn Institute.

Baumgartner, Frank R., and Beth L. Leech. 1998. *Basic Interests: The Importance of Groups in Politics and in Political Science.* Princeton: Princeton University Press.

Baumgartner, Frank R., and Bryan D. Jones. 1993. *Agendas and Instability in American Politics.* Chicago: University of Chicago Press.

Beatley, Timothy. 2000. *Green Urbanism: Learning from the European Cities.* Washington, D.C.: Island.

Bednar, Charles Sokol. 2003. *Transforming the Dream: Ecologism and the Shaping of an Alternative American Vision.* Albany: State University of New York Press.

Belcher, Wyatt Winton. 1947. *The Economic Rivalry Between St. Louis and Chicago, 1850–1880.* New York: Colombia University Press.

Bernstein, Steven. 2001. *The Compromise of Liberal Environmentalism.* New York: Colombia University Press.

Block, Fred. 1987. *Revising State Theory: Essays in Politics and Postindustrialism.* Philadelphia: Temple University Press.

———. 1990. *Postindustrial Possibilities: A Critique of Economic Discourse.* Los Angeles: University of California Press.

Boone, Christopher G., and Ali Modarres. 1999. "Creating a Toxic Neighborhood in Los Angeles County: A Historical Examination of Environmental Inequity." *Urban Affairs Review* 35, no. 2: 163–187.

Borenstein, Seth. 2000 May 10. "Going Nowhere." *Miami Herald*, p. C1.

Boschken, Herman L. 2002. *Social Class, Politics, and Urban Markets: The Makings of Bias in Policy Outcomes.* Stanford: Stanford University Press.

Bottles, Scott. 1987. *Los Angeles and the Automobile: The Making of the Modern City.* Los Angeles: University of California Press.

Bowles, Samuel, David M. Gordon & Thomas E. Weisskopf. 1983. *Beyond the Waste Land: A Democratic Alternative to Economic Decline.* Garden City, N.Y.: Anchor Press/Doubleday.

Brienes, Marvin. 1975. The Fight Against Smog in Los Angeles, 1943–1957. Ph.D. diss., University of California, Davis.

———. 1976. "Smog Comes to Los Angeles." *Southern California Quarterly* 58, no. 4: 515–532.

Brown, Donald A. 2002. *American Heat: Ethical Problems with the United States' Response to Global Warming.* Lanham, Md.: Rowman & Littlefield.

Brownell, Blaine. 1975. *The Urban Ethos in the South.* Baton Rouge: Louisiana State University Press.

Browning, Rufus P., Dale R. Marshall, and David H. Tabb. 1984. *Protest is Not Enough.* Berkeley: University of California Press.

———. 1989. *Racial Politics in American Cities.* New York: Longman.

Bryner, Gary C. 1995. *Blue Skies, Green Politics: The Clean Air Act of 1990 and Its Implementation.* 2nd ed. Washington, D.C.: Congressional Quarterly Press.

———. 1997. "Market Incentives in Air Pollution Control." In *Flash Points in Environmental Policymaking: Controversies in Achieving Sustainability.* Edited by Sheldon Kamieniecki, George A. Gonzalez and Robert O. Vos. Albany: State University of New York Press.

Bullard, Robert D. 1990. *Dumping in Dixie.* Boulder: Westview.

Cackette, Tom, Chief Deputy Executive Officer, California Air Resources Board. 2000. Interview by author, 14 March, Sacramento. Tape Recording.

Cahn, Matthew A. 1995. *Environmental Deceptions: The Tension Between Liberalism and Environmental Policymaking in the United States.* Albany: State University of New York Press.

California Air Resources Board. 2004. *The 2004 California Almanac of Emissions & Air Quality.* Sacramento: California Environmental Protection Agency.

California Council for Environmental & Economic Balance (CCEEB). 1990. *Alternative Motor Vehicle Fuels to Improve Air Quality: Options and Implications for California.* San Francisco: California Council for Environmental & Economic Balance.

———. 2000a. *California Council for Environmental & Economic Balance.* San Francisco: California Council for Environmental & Economic Balance.

———. 2000b. *Mission.* San Francisco: California Council for Environmental & Economic Balance.

Campbell, J. M., Administrative Director, General Motors Corp. 1953 March 26. "To Kenneth Hahn, Los Angeles County Supervisor." In *Smog: A Factual Record of Correspondence between Kenneth Hahn, Los Angeles County Supervisor and the Presidents of General Motors, Ford and Chrysler: 1953–1970.* Los Angeles: Los Angeles County Board of Supervisors.

Carlin, Alan P., and George E. Kocher. 1971. *Environmental Problems: Their Causes, Cures, and Evolution: Using Southern California Smog as an Example.* Santa Monica, Calif.: Rand Corporation.

Carmichael, Tim, Executive Director, Coalition for Clean Air. 2000. Interview by author, 15 March, San Francisco. Tape recording.

Carpenter, Daniel P. 2001. *The Forging of Bureaucratic Autonomy: Reputations, Networks, and Policy Innovations in Executive Agencies, 1862–1928.* Princeton: Princeton University Press.

Carter, Neil. 2001. *The Politics of the Environment: Ideas, Activism, Policy.* Cambridge, U.K.: Cambridge University Press.

Casner, Nicholas. 1999. "Polluter versus Polluter: The Pennsylvania Railroad and the Manufacturing of Pollution Policies in the 1920s." *Journal of Policy History* 11, no. 2: 179–200.

Chamber of Commerce of Pittsburgh. 1900. *Yearbook and Directory of the Chamber of Commerce of Pittsburgh.* Pittsburgh: Chamber of Commerce of Pittsburgh.

Chandler, Alfred, Jr. 1972. "Anthracite Coal and the Beginnings of the Industrial Revolution in the United States." *Business History Review* 46 (summer): 141–181.

Chayne, Charles A., Vice President, General Motors Corp. 1954 Nov. 2. "To Kenneth Hahn, Los Angeles County Supervisor." In *Smog: A Factual Record of Correspondence between Kenneth Hahn, Los Angeles County Supervisor and the Presidents of General Motors, Ford and Chrysler: 1953–1970.* Los Angeles: Los Angeles County Board of Supervisors.

Cheape, Charles W. 1980. *Moving the Masses: Urban Public Transit in New York, Boston, and Philadelphia, 1880–1912.* Cambridge: Harvard University Press.

Cherni, Judith A. 2002. *Economic Growth versus the Environment: The Politics of Wealth, Health and Air Pollution.* New York: Palgrave.

Chicago Association of Commerce. 1915. *Smoke Abatement and Electrification of Railway Terminals in Chicago.* Chicago: McNally.

Christianson, Gale E. 1999. *Greenhouse: The 200–Year Story of Global Warming.* New York: Walker.

Christoff, Peter. 1996. "Ecological Modernization, Ecological Modernities." *Environmental Politics* 5, no. 3: 476–500.

Clawson, Dan, Alan Neustadtl, and Mark Weller. 1998. *Dollars and Votes: How Business Campaign Contributions Subvert Democracy.* Philadelphia: Temple University Press.

Coase, R. H. 1960. "The Problem of Social Cost." *Journal of Law and Economics* 3 (October): 1–44.

Cohen, Jean L., and Andrew Arato. 1992. *Civil Society and Political Theory.* Cambridge: MIT Press.

Cohen, Michael P. 1988. *The History of the Sierra Club, 1892–1970.* San Francisco: Sierra Club Books.

Cohen, Richard E. 1995. *Washington at Work: Back Rooms and Clean Air.* Boston: Allyn and Bacon.

Cole, Luke W., and Sheila R. Foster. 2001. *From the Ground Up: Environmental Racism and the Rise of the Environmental Justice Movement*. New York: New York University Press.

"Colorado's High-Oxygen Fuel Test Runs Smoothly." 1988 March 1. *New York Times*, p. B5.

Commons, John R. 1924. *Legal Foundations of Capitalism*. New York: Macmillan.

Cone, Marla. 1995 Dec. 7. "State Offers to Delay Electric Car Mandate." *Los Angeles Times*, p. A3.

———. 1995 Dec. 20. "Air Panel Bending Under Pressure." *Los Angeles Times*, p. A3.

———. 1995 Dec. 22. "State Panel Puts Electric Car Mandate in Reverse." *Los Angeles Times*, p. A1.

———. 1998 July 18. "Air Board Seeks Tighter Auto Emission Limits." *Los Angeles Times*, p. A1.

———. 1999 Oct. 30. "Vehicles Blamed for a Greater Share of Smog." *Los Angeles Times*, p. A1.

Conot, Robert. 1986. *American Odyssey: A History of a Great City*. Detroit: Wayne State University Press.

Crawford, David F. 1913. "The Abatement of Locomotive Smoke." *Industrial World* 47, no. 2: 1095–1100

Crenson, Matthew A. 1971. *The Un-Politics of Air Pollution*. Baltimore: Johns Hopkins University Press.

Cronon, William. 1991. *Nature's Metropolis: Chicago and the Great West*. New York: Norton.

Crowley, Kate. 1999. "Jobs and Environment: The 'Double Dividend' of Ecological Modernization." *International Journal of Social Economics* 26, no. 7/8/9: 1013–1026.

Crump, Spencer. 1988. *Ride the Big Red Cars: The Pacific Electric Story*. 7th ed. Glendale, Calif.: Trans-Anglo.

Curcio, Vincent. 2000. *Chrysler: The Life and Times of an Automotive Genius*. New York: Oxford University Press.

Cushman, John H., Jr. 1998 June 7. "E.P.A. and States Found to be Lax on Pollution Law." *New York Times*, p. 1.

———. 1998 August 5. "Governors Demand a Larger Voice in Clean-Water Programs." *New York Times*, p. A20.

———. 2002 July 2. "California Lawmakers Vote to Lower Auto Emissions." *New York Times*, p. A14.

Cyphers, Christopher J. 2002. *The National Civic Federation and the Making of New Liberalism, 1900–1915*. Westport, Conn.: Praeger.

Dahl, Robert A., and Charles E. Lindblom. 1953. *Politics, Economics, and Welfare*. New Haven: Yale University Press.

Dahl, Robert A. 1956. *A Preface to Democratic Theory*. Chicago: University of Chicago Press.

———. 1958. "A Critique of the Ruling Elite Model." *American Political Science Review* 52, no. 2: 463–469.

———. 1959. "Business and Politics: A Critical Appraisal of Political Science." *American Political Science Review* 53, no. 1: 1–34.

———. 1961. *Who Governs?: Democracy and Power in an American City*. New Haven: Yale University Press.

Davis, Charles, and Sandra K. Davis. 1999. "State Enforcement of the Federal Hazardous Waste Program." *Polity* 31, no. 3: 451–68.

Davis, David. 1993. *Energy Politics*. New York: St. Martin's Press.

Davis, Devra. 2002. *When Smoke Ran Like Water: Tales of Environmental Deception and the Battle Against Pollution*. New York: Basic.

Davis, Donald F. 1979. "Mass Transit and Private Ownership: An Alternative Perspective on the Case of Toronto." *Urban History Review* 3 (Feb.): 60–98.

Davison, Aidan. 2001. *Technology and the Contested Meanings of Sustainability*. Albany: State University of New York Press.

Dawson, Bill. 1999 Nov. 20. "State Points to Chevron Plant Emissions in Day that puts Houston No. 1 in Smog." *Houston Chronicle*, p. A1.

DeLeon, Richard Edward. 1992. *Left Coast City: Progressive Politics in San Francisco, 1975–1991*. Lawrence: University Press of Kansas.

Derrick, Peter. 2001. *Tunneling to the Future: The Story of the Great Subway Expansion That Saved New York*. New York: New York University.

Desai, Uday, ed. 2002. *Environmental Politics and Policy in Industrialized Countries*. Cambridge: MIT Press.

Desfor, Gene, and Roger Keil. 2004. *Nature and the City: Making Environmental Policy in Toronto and Los Angeles*. Tucson: University of Arizona Press.

Dewees, Donald N. 1970. "The Decline of the American Street Railways." *Traffic Regulation* 24: 563–81.

Dewey, Scott. 2000. *Don't Breathe the Air: Air Pollution and U.S. Environmental Politics, 1945–1970*. College Station, Tex.: Texas A&M University Press.

Doherty, Brian. 2002. *Ideas and Actions in the Green Movement*. New York: Routledge.

Domhoff, G. William. 1967. *Who Rules America?* Englewood Cliffs, N.J.: Prentice Hall.

———. 1970. *The Higher Circles: The Governing Class in America*. New York: Vintage.

———. 1974. *The Bohemian Grove and Other Retreats*. New York: Harper and Row.

———. 1978a. *The Powers that Be*. New York: Random House.

———. 1978b. *Who Really Rules: New Haven and Community Power Reexamined*. Santa Monica, CA: Goodyear.

———. 1990. *The Power Elite and the State*. New York: Aldine de Gruyter.

———. 1996. *State Autonomy or Class Dominance?: Case Studies on Policymaking in America*. New York: Aldine de Gruyter.

———. 2002. *Who Rules America? Power and Politics.* 4th ed. New York: McGraw-Hill.

Dowie, Mark. 1995. *Losing Ground: American Environmentalism at the Close of the Twentieth Century.* Cambridge: MIT Press.

———. 2001. *American Foundations: An Investigative History.* Cambridge: MIT Press.

Dreier, Peter, John Mollenkopf, and Todd Swanstrom. 2001. *Place Matters: Metropolitics for the Twenty-First Century.* Lawrence: University Press of Kansas.

Dryzek, John S. 1987. *Rational Ecology: Environment and Political Economy.* New York: Blackwell.

———. 1996a. "Political Inclusion and the Dynamics of Democratization." *American Political Science Review* 90, no. 1: 475–487.

———. 1996b. *Democracy in Capitalist Times.* New York: Oxford University Press.

———. 1997. *The Politics of the Earth.* New York: Oxford University Press.

Dryzek, John S., David Downs, Christian Hunold, and David Schlosberg, with Hans-Kristian Hernes. 2003. *Green States and Social Movements: Environmentalism in the United States, United Kingdom, Germany, and Norway.* New York: Oxford University Press.

Eakins, David. 1969. "Business Planners and America's Postwar Expansion." In *Corporations and the Cold War.* Edited by David Horowitz. New York: Monthly Review Press.

———. 1972. "Policy-Planning for the Establishment." In *A New History of Leviathan.* Edited by Ronald Radosh and Murray N. Rothbard. New York: E. P. Dutton.

Edelman, Murray. 1964. *The Symbolic Uses of Politics.* Urbana, Ill.: University of Illinois Press.

———. 1977. *Political Language.* New York: Academic Press.

———. 1988. *Constructing the Political Spectacle.* Chicago: University of Chicago Press.

Eisinger, Peter K. 1988. *The Rise of the Entrepreneurial State: State and Local Economic Development Policy in the United States.* Madison: University of Wisconsin Press.

Elkin, Stephen L. 1987. *City and Regime in the American Republic.* Chicago: University of Chicago Press.

Ellerman, A. Denny, Paul L. Joskow, Richard Schmalensee, Juan-Pablo Montero, and Elizabeth M. Bailey. 2000. *Markets for Clean Air: The U.S. Acid Rain Program.* Cambridge, U.K.: Cambridge University Press.

Ely, Richard T. 1914. *Property and Contract in Their Relations to the Distribution of Wealth.* Vols. 1–2. New York: Macmillan.

Engler, Robert. 1961. *The Politics of Oil: A Study of Private Power and Democratic Directions.* New York: Macmillan.

———. 1977. *The Brotherhood of Oil: Energy Policy and the Public Interest.* Chicago: University of Chicago Press.

Erie, Steven P. 2004. *Globalizing L.A.: Trade, Infrastructure, Regional Development.* Stanford: Stanford University Press.

Fainstein, Susan S. 2001. *The City Builders: Property Development in New York and London, 1980–2000.* Lawrence: University Press of Kansas.

Farber, David. 2002. *Sloan Rules: Alfred P. Sloan and the Triumph of General Motors.* Chicago: University of Chicago Press.

Fawcett, Jeffry. 1990. *The Political Economy of Smog in Southern California.* New York: Garland.

Feitelson, Eran, and Erik T. Verhoef. 2001. *Transport and Environment: In Search of Sustainable Solutions.* Cheltenham, U.K.: Edward Elgar.

Fellmeth, Robert C. 1973. *Politics of Land: Ralph Nader's Study Group Report on Land Use in California.* New York: Grossman.

Ferman, Barbara. 1996. *Challenging the Growth Machine: Neighborhood Politics in Chicago and Pittsburgh.* Lawrence: University Press of Kansas.

Finegold, Kenneth, and Theda Skocpol. 1995. *State and Party in America's New Deal.* Madison: University of Wisconsin Press.

Firor, John, and Judith Jacobsen. 2002. *The Crowded Greenhouse: Population, Climate Change, and Creating a Sustainable World.* New Haven: Yale University Press.

Fisher, Peter S., and Alan H. Peters. 1998. *Industrial Incentives: Competition Among American States and Cities.* Kalamazoo, Mich.: W. E. Upjohn Institute.

Fishman, Robert. 1987. *Bourgeois Utopias: The Rise and Fall of Suburbia.* New York: Basic.

Flanagan, Maureen A. 1996. "The City Profitable, the City Livable." *Journal of Urban History* 22, no. 2: 163–190.

Flink, James. 1975. *The Car Culture.* Cambridge: MIT Press.

———. 1990. *The Automobile Age.* Cambridge: MIT Press.

Fogelson, Robert M. 1967. *The Fragmented Metropolis, Los Angeles, 1850–1930.* Cambridge: Harvard University Press.

———. 2001. *Downtown: Its Rise and Fall, 1880–1950.* New Haven: Yale University Press.

Foley, Duncan K. 2003. *Unholy Trinity: Labor, Capital, and Land in the New Economy.* New York: Routledge.

Ford, Henry, II, President, Ford Motor Co. 1953 Feb. 19. "To Kenneth Hahn, Los Angeles County Supervisor." In *Smog: A Factual Record of Correspondence between Kenneth Hahn, Los Angeles County Supervisor and the Presidents of General Motors, Ford and Chrysler: 1953–1970.* Los Angeles: Los Angeles County Board of Supervisors.

Ford, Richard C. 2001. "The Political Response to Black Insurgency: A Critical Test of Competing Theories of the State." *American Political Science Review* 95, no. 1: 115–130.

Foss, Phillip O. 1960. *Politics and Grass.* Seattle: University of Washington.

Foster, Mark S. 1971. The Decentralization of Los Angeles During the 1920s. Ph.D. diss., University of Southern California.

————. 1975. "The Model-T, the Hard Sell, and Los Angeles's Urban Growth: The Decentralization of Los Angeles during the 1920s." *Pacific Historical Review* 44 (Nov.): 459–484.

————. 1981. *From Streetcar to Superhighway: American City Planners and Urban Transportation, 1900–1940.* Philadelphia: Temple University Press.

————. 1992. "The Role of the Automobile in Shaping a Unique City: Another Look." In *The Car and the City: The Automobile, the Built Environment, and Daily Urban Life.* Edited by Martin Wachs and Margaret Crawford. Ann Arbor: University of Michigan Press.

Fulton, William. 2001. *The Reluctant Metropolis: The Politics of Urban Growth in Los Angeles.* Baltimore: Johns Hopkins University Press.

Gainsborough, Juliet F. 2001. "Bridging the City-Suburb Divide: States and the Politics of Regional Cooperation." *Journal of Urban Affairs* 23, no. 5: 497–512.

————. 2002. "Slow Growth and Urban Sprawl: Support for a New Regional Agenda?" *Urban Affairs Review* 37, no. 5: 728–744.

————. 2003. "Business Organizations as Regional Actors: The Politics of Regional Cooperation in Metropolitan America." *Polity* 35, no. 4: 555–572.

Game, Kingsley. 1979. "Controlling Air Pollution: Why Some States Try Harder." *Policy Studies Journal* 7: 728–738.

Gonzalez, George A. 1998. "The Conservation Policy Network, 1890–1910: The Development and Implementation of 'Practical' Forestry." *Polity* 31, no. 2: 269–299.

————. 1999 June. "Book Review of *Pluralism-by-the-Rules* by Edward P. Weber." *American Political Science Review* 93, no. 2: 461–462.

————. 2000. "Book Review of *Voices and Echoes for the Environment* by Ronald Shaiko and *Eco-Wars* by Ronald Libby." *American Political Science Review* 94, no. 4: 950–951.

————. 2001a. *Corporate Power and the Environment: The Political Economy of U.S. Environmental Policy.* Lanham, Md.: Rowman & Littlefield.

————. 2001b. "Ideas and State Capacity, or Business Dominance? A Historical Analysis of Grazing on the Public Grasslands." *Studies in American Political Development* 15 (fall): 234–244.

Gordon, Colin. 1994. *New Deals: Business, Labor, and Politics in America, 1920–1935.* Cambridge, U.K.: Cambridge University Press.

Gorman, Hugh, and Barry D. Solomon. 2002. "The Origins, Practice, and Limits of Emissions Trading." *Journal of Policy History* 14, no. 3: 293–320.

Gorz, Andre. 1985. *Ecology As Politics.* Boulder: Westview.

————. 1994. *Capitalism, Socialism, Ecology.* New York: Norton.

Gottlieb, Robert, and Irene Wolt. 1977. *Thinking Big: The Story of the Los Angeles Times Its Publishers and Their Influence on Southern California.* New York: G. P. Putnam's Sons.

Gough, Ian. 2000. *Global Capital, Human Needs and Social Policies*. New York: St. Martin's.

Gould, Kenneth A., Allan Schnaiberg, and Adam S. Weinberg. 1996. *Local Environmental Struggles: Citizen Activism in the Treadmill of Production*. New York: Cambridge University Press.

Graham, Otis L., ed. 2000. *Environmental Policy and Politics, 1960s-1990s*. University Park, Pa.: Pennsylvania State University Press.

Grant, Wyn. 1996. *Autos, Smog, and Pollution Control*. Brookfield, Vt.: Edward Elgar.

Greer, Edward. 1974. "Air Pollution and Corporate Power: Municipal Reform Limits in a Black City." *Politics and Society* 4, no. 4: 483–510.

Grinder, R. Dale. 1978. "The Smoke Abatement Campaign in Pittsburgh Before World War I." *The Western Pennsylvania Historical Magazine* 61, no. 3: 187–202.

———. 1980. "The Battle for Clean Air: The Smoke Problem in Post–Civil War America." In *Pollution and Reform in American Cities, 1870–1930*. Edited by Martin V. Melosi. Austin, Tex.: University of Texas Press.

Grossman, Gene M., and Elhanan Helpman. 2001. *Special Interest Politics*. Cambridge: MIT Press.

Guber, Deborah Lynn. 2003. *The Grassroots of a Green Revolution: Polling America on the Environment*. Cambridge: MIT Press.

Gugliotta, Angela. 2000. "Class, Gender, and Coal Smoke: Gender Ideology and Environmental Injustice in Pittsburgh, 1868–1914." *Environmental History* 5, no. 2: 165–193.

———. 2003. "How, When, and for Whom Was Smoke a Problem in Pittsburgh?" In *Devastation and Renewal: An Environmental History of Pittsburgh and its Region*. Edited by Joel A. Tarr. Pittsburgh: University of Pittsburgh Press.

Hajer, Maarten A. 1995. *The Politics of Environmental Discourse*. New York: Oxford University Press.

Hakim, Danny. 2003 April 25. "California Regulators Modify Auto Emissions Mandate." *New York Times*, p. A24.

———. 2003 August 12. "Automakers Drop Suits on Air Rules." *New York Times*, p. A1.

Hall, Peter. 1995. "A European Perspective on the Spatial Links between Land Use, Development and Transport." In *Transport and Urban Development*. Edited by David Banister. London: E & F Spon.

Harvey, David. 1985. *The Urbanization of Capital: Studies in the History and Theory of Capitalist Urbanization*. Baltimore: Johns Hopkins University Press.

Hathaway, Janet, Senior Attorney, Natural Resources Defense Council. 2000. Interview by author, 16 March, San Francisco. Tape Recording.

Hays, Samuel. 1964. "The Politics of Reform in Municipal Government in the Progressive Era." *Pacific Northwest Quarterly* 55, no. 4: 157–169.

———. 1987. *Beauty, Health, and Permanence: Environmental Politics in the United States, 1955–1985*. Cambridge: Cambridge University Press.

———. 2000. *A History of Environmental Politics Since 1945*. Pittsburgh: University of Pittsburgh Press.

Hayward, Clarissa Rile. 2000. *De-Facing Power*. New York: Cambridge University Press.

Heclo, Hugh. 1978. "Issue Networks and the Executive Establishment." In *The New American Political System*. Edited by Anthony King. Washington, D.C.: American Enterprise Institute for Public Policy Research.

Heinz, John P., Edward O. Laumann, Robert L. Nelson, and Robert H. Salisbury. 1993. *The Hollow Core: Private Interests in National Policymaking*. Cambridge: Harvard University Press.

Herbert, Josef H. 2004 June 30. "EPA Says Counties in 22 States Have Bad Air Because of Soot." Associated Press.

Hernandez, Ramona. 2002. *The Mobility of Labor under Advanced Capitalism: Dominican Migration to the United States*. New York: Columbia University Press.

Higgins-Evenson, R. Rudy. 2003. *The Price of Progress: Public Services, Taxation, and the American Corporate State, 1877 to 1929*. Baltimore: Johns Hopkins University Press.

Hirt, Paul W. 1994. *A Conspiracy of Optimism: Management of the National Forests since World War Two*. Lincoln, Nebr.: University of Nebraska Press.

Hise, Greg. 1997. *Magnetic Los Angeles: Planning the Twentieth-Century Metropolis*. Baltimore: Johns Hopkins University Press.

———. 2001. "'Nature's Workshop' Industry and Urban Expansion in Southern California, 1900–1950." *Journal of Historical Geography* 27, no. 1: 74–92.

Hornborg, Alf. 2001. *The Power of the Machine: Global Inequalities of Economy, Technology, and Environment*. Walnut Creek, Calif.: Altamira.

Hoyt, Homer. 1933. *One Hundred Years of Land Values in Chicago: The Relationship of the Growth of Chicago to the Rise in its Land Values, 1830–1933*. Chicago: University of Chicago Press.

Hulse, Carl. 2004 April 3. "House Approves Highway Bill, Raising Prospect of Veto Showdown With Bush." *New York Times*, p. A8.

Hurley, Andrew. 1995. *Class, Race, and Industrial Pollution in Gary, Indiana, 1945–1980*. Chapel Hill, NC: University of North Carolina Press.

Inglehart, Ronald. 1977. *The Silent Revolution: Changing Values & Political Styles among Western Publics*. Princeton: Princeton University Press.

Isaac, Jeffrey. 1993. "Civil Society and the Spirit of Revolt." *Dissent* 40 (summer): 356–361.

———. 2003. *The Poverty of Progressivism: The Future of American Democracy in a Time of Liberal Decline*. Lanham, Md. Rowman & Littlefield.

Jackson, Kenneth T. 1985. *Crabgrass Frontier: The Suburbanization of the United States*. New York: Oxford University Press.

Jacobson, Mark Z. 2002. *Atmospheric Pollution: History, Science, and Regulation*. New York: Cambridge University Press.

Jaher, Frederic Cople. 1982. *The Urban Establishment: Upper Strata in Boston, New York, Charleston, Chicago, and Los Angeles*. Urbana, Ill.: University Press of Illinois.

Jonas, Andrew E. G., and David Wilson, eds. 1999. *The Urban Growth Machine: Critical Perspectives Two Decades Later*. Albany: State University of New York Press.

Jones, Charles O. 1975. *Clean Air*. Pittsburgh: University of Pittsburgh Press.

Jones, Holway R. 1965. *John Muir and the Sierra Club: The Battle for Yosemite*. San Francisco: Sierra Club.

Kamieniecki, Sheldon, and Michael Farrell. 1991. "Intergovernmental Relations and Clean-Air Policy in Southern California." *Publius* 21, no. 3: 143–154.

Kamieniecki, Sheldon, David Shafie, and Julie Silvers. 1999. "Forming Partnerships in Environmental Policy." *American Behavioral Scientist* 43, no. 1: 107–123.

Kay, Jane Holtz. 1998. *Asphalt Nation: How the Automobile Took over America and How We Can Take It Back*. Berkeley: University of California Press.

Keating, Ann Durkin. 1988. *Building Chicago: Suburban Developers and the Creation of a Divided Metropolis*. Columbus, Ohio: Ohio State University Press.

Kemp, Kathleen. 1981. "Symbolic and Strict Regulation in the American States." *Social Science Quarterly* 62, no. 3: 516–526.

Kennedy, Harold W. 1954. *The History, Legal and Administrative Aspects of Air Pollution Control in the County of Los Angeles*. Los Angeles: Board of Supervisors of the County of Los Angeles.

Kennedy, Randy. 2003 April 20. "The Day The Traffic Disappeared." *New York Times*, sec. 6 , p. 42.

Kenworthy, Jeffrey R., and Felix B. Laube, with Peter Newman, Paul Barter, Tamim Raad, Chamlong Poboon, and Benedicto Guia, Jr. 1999. *An International Sourcebook of Automobile Dependence in Cities 1960–1990*. Boulder: University Press of Colorado.

Klyza, Christopher. 1996. *Who Controls the Public Lands?: Mining, Forestry, and Grazing Policies, 1870–1990*. Chapel Hill, N.C.: University of North Carolina Press.

Kohn, Margaret. 2003. *Radical Space: Building the House of the People*. Ithaca, N.Y.: Cornell University Press.

Kolko, Gabriel. 1977. *The Triumph of Conservatism: A Reinterpretation of American History, 1900–1916*. New York: Free Press. Originally published in 1963.

Kozlowski, Paul J., and James K. Weekly. 1990. "Explaining Interstate Variations in Foreign Direct Investment in the United States." *Regional Science Perspectives* 20, no. 2: 3–25.

Kraft, Michael E. 1993. "Air Pollution in the West: Testing the Limits of Public Support with Southern California's Clean Air Policy." In *Environmental Politics and Policy in the West*. Edited by Zachary Smith. Dubuque, Iowa: Kendall/Hunt.

———. 1994. "Environmental Gridlock: Searching for Consensus in Congress." In *Environmental Policy in the 1990s*. 2nd ed. Edited by N. J. Vig and M. E. Kraft. Washington, D.C.: Congressional Quarterly Press.

———. 2001. *Environmental Policy and Politics.* 2nd ed. New York: Addison Wesley Longman.

———. 2002. "Environmental Policy and Politics in the United States: Toward Environmental Sustainability?" In *Environmental Politics and Policy in Industrialized Countries.* Edited by Uday Desai. Cambridge: MIT Press.

Krier, James E., and Edmund Ursin 1977. *Pollution and Policy: A Case Essay on California and Federal Experience with Motor Vehicle Air Pollution.* Los Angeles: University of California Press.

Laird, Frank N. 2001. *Solar Energy, Technology Policy, and Institutional Values.* Cambridge, U.K.: Cambridge University Press.

Lamare, James W. 1993. *California Politics.* New York: West.

———. 2000. *Texas Politics: Economics, Power, and Policy.* 7th ed. Belmont, Calif.: Wadsworth.

Lane, Julia, Dennis Clennon, and James McCabe. 1989. "Measures of Local Business Climate: Alternative Approaches." *Regional Science Perspectives* 19, no. 1: 89–100.

Lange, Leif Erik. 1999. "Transportation and Environmental Costs of Auto Dependency." In *California's Threatened Environment.* Edited by Tim Palmer. Washington, D.C.: Island.

Lee, Gary. 1996. "Compromising on Clean Air Act: Under Republican Pressure, EPA Reduces Enforcement Efforts." *Washington Post*, p. A1.

Lee, Jennifer 8. 2004 April 13. "Clear Skies No More for Millions as Pollution Rule Expands." *New York Times*, p. A16.

Leggett, Jeremy. 2001. *The Carbon War: Global Warming and the End of the Oil Era.* New York: Routledge.

Lindblom, Charles E. 1977. *Politics and Markets: The World's Political-Economic Systems.* New York: Basic Books.

———. 1982. "The Market as Prison." *Journal of Politics* 44, no. 2: 324–336.

Lippman, Thomas W. 1990 June 12. "Gasoline Formula Fuels Controversy." *Washington Post*, p. D1.

Lisowski, Michael. 2002. "Playing the Two-level Game: U.S. President Bush's Decision to Repudiate the Kyoto Protocol." *Environmental Politics* 11, no. 4: 101–119.

Litfin, Karen. 1994. *Ozone Discourses: Science and Politics in Global Environmental Cooperation.* New York: Columbia University Press.

Logan, John R., and Harvey L. Molotch. 1987. *Urban Fortunes: The Political Economy of Place.* Berkeley: University of California Press.

Logan, Michael. 1995. *Fighting Sprawl and City Hall: Resistance to Urban Growth in the Southwest.* Tucson: University of Arizona Press.

Lotchin, Roger W. 1992. *Fortress California, 1910–1960.* New York: Oxford University Press.

Lowery, David, and Virginia Gray. 2004. "Review Essay: A Neopluralist Perspective on Research on Organized Interests." *Political Research Quarterly* 57, no. 1: 163–175.

Lowi, Theodore J. 1979. *The End of Liberalism: The Second Republic of the United States.* New York: Norton.

Lowry, William R. 1992. *The Dimensions of Federalism: State Governments and Pollution Control Policies.* Durham, N.C.: Duke University Press.

Lucas, Robert W., Consultant for California Council for Environmental & Economic Balance. 2000. Telephone interview by author, 30 March, Sacramento.

Luger, Stan. 2000. *Corporate Power, American Democracy, and the Automobile Industry.* Cambridge, U.K.: Cambridge University Press.

Luke, Timothy W. 1997. *Ecocritique: Contesting the Politics of Nature, Economy, and Culture.* Minneapolis: University of Minnesota Press.

Lukes, Steven. 1974. *Power: A Radical View.* London: Macmillan.

"Lung Association Faults U.S. on Enforcing Clean Air Laws." 2002 May 2. *New York Times,* p. A24.

Manley, John F. 1983. "Neo-Pluralism: a Class Analysis of Pluralism I and Pluralism II." *American Political Science Review* 77, no. 2: 368–383.

Mark, Jason, Transportation Analyst, Union of Concerned Scientists. 2000. Interview by author, 13 March, San Francisco. Tape recording.

Marzotto, Toni, Vicky Moshier Burnor, and Gordon Scott Bonham. 2000. *The Evolution of Public Policy: Cars and the Environment.* Boulder: Lynne Rienner.

Mayer, Harold M. and Richard Wade. 1969. *Chicago: Growth of a Metropolis.* Chicago: University of Chicago Press.

Mazmanian, Daniel A. 1999. "Los Angeles' Transition from Command-and-Command to Market-Based Clear Air Policy Strategies and Implementation." In *Toward Sustainable Communities.* Edited by Daniel A. Mazmanian and Michael Kraft. Cambridge: MIT Press.

McConnell, Grant. 1966. *Private Power and American Democracy.* New York: Knopf.

McDougal, Dennis. 2001. *Privileged Son: Otis Chandler and the Rise and Fall of the L.A. Times Dynasty.* New York: Perseus.

McFarland, Andrew S. 1987. "Interest Groups and Theories of Power in America." *British Journal of Political Science* 17, no. 2: 129–147.

———. 1993. *Cooperative Pluralism: The National Coal Policy Experiment.* Lawrence: University Press of Kansas.

———. 2004. *Neopluralism: The Evolution of Political Process Theory.* Lawrence: University Press of Kansas.

McKay, John P. 1976. *Tramways and Trolleys: The Rise of Urban Mass Transit in Europe.* Princeton: Princeton University Press.

———. 1988. "Comparative Perspectives on Transit in Europe and the United States, 1850–1914." In *Technology and the Rise of the Networked City in Europe and America*. Edited by Joel A. Tarr and Gabriel Dupuy. Philadelphia: Temple University Press.

McShane, Clay. 1974. *Technology and Reform: Street Railways and the Growth of Milwaukee, 1887–1900*. Madison: State Historical Society of Wisconsin.

———. 1988. "Urban Pathways: The Street and Highway, 1900–1940." In *Technology and the Rise of the Networked City in Europe and America*. Edited by Joel A. Tarr and Gabriel Dupuy. Philadelphia: Temple University Press.

———. 1994. *Down the Asphalt Path: The Automobile and the American City*. New York: Columbia University Press.

Melosi, Martin V. 2001. *Effluent America: Cities, Industry, Energy, and the Environment*. Pittsburgh: University of Pittsburgh Press.

Milbrath, Lester W. 1989. *Envisioning a Sustainable Society: Learning Our Way Out*. Albany: State University of New York Press.

———. 1995. "Psychological, Cultural, and Informational Barriers to Sustainability." *Journal of Social Issues* 51, no.4: 101–120.

———. 1996. *Learning to Think Environmentally*. Albany: State University of New York Press.

Miliband, Ralph. 1969. *The State in Capitalist Society*. New York: Basic.

Mintz, Beth and Michael Schwartz. 1985. *The Power Structure of American Business*. Chicago: University of Chicago Press.

Mintz, Joel A. 1995. *Enforcement at the EPA*. Austin: University of Texas Press.

Mitchell, Timothy. 1991. "The Limits of the State: Beyond Statist Approaches and Their Critics." *American Political Science Review* 85, no. 1: 77–96.

Moehring, Eugene P. 2004. *Urbanism and Empire in the Far West, 1840–1890*. Reno: University of Nevada Press.

Mol, Arthur P. J. 2001. *Globalization and Environmental Reform: The Ecological Modernization of the Global Economy*. Cambridge: MIT Press.

———. 2002. "Ecological Modernization and the Global Economy." *Global Environmental Politics* 2, no. 2:92–115.

Mol, Arthur P. J., and David A. Sonnenfeld. 2000. "Ecological Modernisation Theory in Debate: A Review." *Environmental Politics* 9, no. 1: 17–49.

Mollenkopf, John C. 1983. *The Contested City*. Princeton: Princeton University Press.

Molotch, Harvey. 1976. "The City as a Growth Machine: Towards of Political Economy of Place." *American Journal of Sociology* 82, no. 2: 309–322.

———. 1979. "Capital and Neighborhood in the United States." *Urban Affairs Quarterly* 14, no. 3: 289–312.

Morag-Levine, Noga. 2003. *Chasing the Wind: Regulating Air Pollution in the Common Law State*. Princeton: Princeton University Press.

Muller, Peter. 1981. *Contemporary Suburban America*. Englewood Cliffs, N.J.: Prentice-Hall.

Neumayer, Eric. 2003. *Weak versus Strong Sustainability: Exploring the Limits of Two Opposing Paradigms*. 2nd ed. Northampton, Mass.: Edward Elgar.

Newell, Peter. 2000. *Climate for Change: Non-State Actors and the Global Politics of the Greenhouse*. Cambridge, U.K.: Cambridge University Press.

Newman, Peter, and Jeffrey Kenworthy. 1999. *Sustainability and Cities: Overcoming Automobile Dependence*. Washington, D.C.: Island.

Nice, David C. 1987. *Federalism: The Politics of Intergovernmental Relations*. New York: St. Martin's.

Nivola, Pietro S. 1999. *Laws of the Landscape: How Policies Shape Cities in Europe and America*. Washington, D.C.: Brookings.

Nivola, Pietro S., and Robert W. Crandall. 1995. *The Extra Mile: Rethinking Energy Policy for Automobile Transportation*. Washington, D.C.: Brookings.

Noble, David F. 1977. *America by Design*. New York: Oxford University Press.

Nordhaus, William, and Joseph Boyer. 2000. *Warming the World: Economic Models of Global Warming*. Cambridge: MIT Press.

Nordlinger, Eric A. 1981. *On the Autonomy of the Democratic State*. Cambridge: Harvard University Press.

Norton, Bryan G. 1991. *Toward Unity among Environmentalists*. New York: Oxford University Press.

O'Connor, James. 2002. *The Fiscal Crisis of the State*. New York: Transaction.

O'Connor, Martin, ed. 1994. *Is Capitalism Sustainable?* New York: Guilford.

Offe, Claus. 1974. "Structural Problems of the Capitalist State: Class Rule and the Political System On the Selectiveness of Political Institutions." In *German Political Studies*. Vol. 1. Edited by Klaus von Beyme. Beverly Hills: Sage.

———. 1984. *Contradictions of the Welfare State*. Cambridge: MIT Press.

Olien, Roger M., and Diana Davids Olien. 2000. *Oil and Ideology: The Cultural Creation of the American Petroleum Industry*. Chapel Hill, N.C.: University of North Carolina Press.

Olson, Mancur. 1971. *The Logic of Collective Action: Public Goods and the Theory of Groups*. Cambridge: Harvard University Press.

Orsi, Richard J. 1985. "'Wilderness Saint' and 'Robber Baron': The Anomalous Partnership of John Muir and the Southern Pacific Company for Preservation of Yosemite National Park." *Pacific Historian* 29 (summer-fall): 136–152.

Patton, Gary A. 1999. "Land Use and Growth Management: The Transformation of Paradise." In *California's Threatened Environment*. Edited by Tim Palmer. Washington, D.C.: Island.

Perelman, Michael. 2003. *The Perverse Economy*. New York: Palgrave Macmillian.

Perez-Pena, Richard. 1999 Nov. 7. "Pataki to Impose Strict New Limits on Auto Emissions." *New York Times*, p. A1.

Peterson, Paul E. 1981. *City limits*. Chicago: University of Chicago Press.

Pincetl, Stephanie. 1999. *Transforming California, A Political History of Land Use and Development*. Baltimore: Johns Hopkins University Press.

Pinderhughes, Raquel. 2004. *Alternative Urban Futures: Planning for Sustainable Development in Cities throughout the World*. Lanham, Md: Rowman & Littlefield.

Piven, Frances, and Richard Cloward. 1971. *Regulating the Poor*. New York: Random House.

Platt, Harold. 1995. "Invisible Gases: Smoke, Gender, and the Redefinition of Environmental Policy in Chicago, 1900–1920." *Planning Perspectives* 10, no. 1: 67–97.

Pollack, Andrew. 2000 Dec. 9. "New Plan Would Scale Back Quota for Electric Cars in California." *New York Times*, p. A21.

Potoski, Matthew. 2001. "Clean Air Federalism: Do States Race to the Bottom?" *Public Administration Review* 61, no. 3: 335–342.

Poulantzas, Nicos. 1973. *Political Power and Social Classes*. London: New Left Books.

Pred, Allan 1966. *The Spatial Dynamics of U.S. Urban-Industrial Growth, 1800–1914*. Cambridge: MIT Press.

———. 1980. *Urban Growth and City-Systems in the United States, 1840–1860*. Cambridge: Harvard University Press.

Preston, Howard L. 1979. *Automobile Age Atlanta: The Making of a Southern Metropolis, 1900–1935*. Athens, Ga.: University of Georgia Press.

———. 1991. *Dirt Roads to Dixie: Accessibility and Modernization in the South, 1885–1935*. Knoxville: University of Tennessee Press.

"Public Called on to Block Crippling of Anti-Smog Bill—Powerful Groups Allied to Punch Holes in Measure." 1947 May 18. *Los Angeles Times*, pt. 1, p. 3.

Pulido, Laura. 1996. *Environmentalism and Economic Justice: Two Chicano Struggles in the Southwest*. Tucson: University of Arizona Press.

Purdum, Todd. 2000 February 13. "Los Angeles Sprawl Bumps Angry Neighbor." *New York Times*, p. 1.

Radford, Gail. 1996. *Modern Housing for America: Policy Struggles in the New Deal Era*. Chicago: University of Chicago Press.

Rajan, Sudhir. 1996. *The Enigma of Automobility: Democratic Politics and Pollution Control*. Pittsburgh: University of Pittsburgh Press.

Reich, Robert B. 2002. *The Future of Success: Working and Living in the New Economy*. New York: Vintage.

Reitan, Marit. 1998. "Ecological Modernisation and 'Realpolitik': Ideas, Interests and Institutions." *Environmental Politics* 7, no. 2: 1–26.

Revkin, Andrew C. 2000 Nov. 26. "Treaty Talks Fails to Find Consensus in Global Warming." *New York Times*, p. 1.

———. 2001 June 12. "Warming Threat Requires Action Now, Scientists Say." *New York Times*, p. A12.

———. 2002 Feb. 15. "Bush Offers Plan for Voluntary Measures to Limit Gas Emissions." *New York Times*, p. A6.

———. 2002 March 20. "Large Ice Shelf in Antarctica Disintegrates at Great Speed." *New York Times*, p. A13.

———. 2002 June 3. "Climate Changing, U.S. Says in Report." *New York Times*, p. A1.

Rhodes, Edwardo L. 2003. *Environmental Justice in America: A New Paradigm*. Bloomington, Ind.: Indiana University Press.

Rhodes, R. A. W., and D. Marsh. 1992. "Policy Networks in British Politics." In *Policy Networks in British Government*. Edited by R. A. W. Rhodes and D. Marsh. Oxford: Clarendon.

Ricardo, David. 1830. *On the Principles of Political Economy, and Taxation*. Washington, D.C.: J. B. Bell.

Rich, Andrew. 2004. *Think Tanks, Public Policy, and the Politics of Expertise*. New York: Cambridge University Press.

Ridge, John Hiski. 1994. "Deconstructing the Clean Air Act: Examining the Controversy Surrounding Massachusetts's Adoption of the California Low Emission Vehicle Program." *Boston College Environmental Affairs Law Review* 22, no. 1 : 163–199.

Ringquist, Evan. 1993. *Environmental Protection at the State Level: Politics and Progress in Controlling Pollution*. Armonk, N.Y.: M. E. Sharpe.

Robbins, William. 1982. *Lumberjacks and Legislators: Political Economy of the U.S. Lumber Industry, 1890–1941*. College Station, Tex.: Texas A&M University Press.

———. 1994. *Colony and Empire: The Capitalist Transformation of the American West*. Lawrence: University Press of Kansas.

Roberts, Paul. 2004. *The End of Oil: On the Edge of a Perilous New World*. New York: Houghton Mifflin.

Roberts, Thomas R. 1969. Motor Vehicular Air Pollution Control in California: A Case Study in Political Unresponsiveness. Honors thesis, Harvard College.

Roelofs, Joan. 2003. *Foundations and Public Policy: The Mask of Pluralism*. Albany: State University of New York Press.

Rome, Adam. 2001. *The Bulldozer in the Countryside: Suburban Sprawl and the Rise of American Environmentalism*. Cambridge, U.K.: Cambridge University Press.

Rosen, Christine M. 1986. *The Limits of Power: Great Fires and the Process of City Growth in America*. Cambridge, U.K.: Cambridge University Press.

———. 1995. "Businessmen Against Pollution in Late Nineteenth Century Chicago." *Business History Review* 69, no. 3: 351–397.

Rosenbaum, Walter A. 1998. *Environmental Politics and Policy*. 4th ed. Washington, D.C.: Congressional Quarterly Press.

Roy, William. 1997. *Socializing Capital: The Rise of the Large Industrial Corporation in America*. Princeton: Princeton University Press.

Runte, Alfred. 1997. *National Parks: The American Experience.* 3d ed. Lincoln, Nebr.: University of Nebraska Press.

Sabatier, Paul A. 1987. "Knowledge, Policy-Oriented Learning, and Policy Change." *Knowledge: Creation, Diffusion, Utilization* 8, no. 4: 649–692.

———. 1999. *Theories of the Policy Process.* Boulder: Westview.

Sanders, M. Elizabeth. 1981. *The Regulation of Natural Gas: Policy and Politics, 1938–1978.* Philadelphia: Temple University Press.

Savitch, H. V., and Paul Kantor. 2002. *Cities in the International Marketplace: The Political Economy of Urban Development in North America and Western Europe.* Princeton: Princeton University Press.

Saward, Michael. 1992. *Co-optive Politics and State Legitimacy.* Dartmouth: Aldershot.

Scheberle, Denise. 1997. *Trust and the Politics of Implementation: Federalism and Environmental Policy.* Washington, D.C.: Georgetown University Press.

Scheible, Michael, Deputy Executive Officer, California Air Resources Board. 2000. Interview by Frank Janeczek, 15 March, Sacramento. Tape recording.

Schlosberg, David. 1999. *Environmental Justice and the New Pluralism.* New York: Oxford University Press.

Schlozman, Kay L., and John T. Tierney. 1986. *Organized Interests and American Democracy.* New York: Harper & Row.

Schnattschneider, E. E. 1960. *The Semisovereign People.* New York: Holt, Rinehart and Winston.

Schrepfer, Susan R. 1983. *The Fight to Save the Redwoods: A History of Environmental Reform, 1917–1978.* Madison: University of Wisconsin Press.

Schultz, Stanley K. 1989. *Constructing Urban Culture: American Cities and City Planning, 1800–1920.* Philadelphia: Temple University Press.

Seelye, Katharine Q. 2001 Sept. 5. "E.P.A. Faults Ohio Agency Headed by a Bush Nominee." *New York Times,* p. A12.

———. 2002 July 1. "Bush Slashing Aid for E.P.A. Cleanup at 33 Toxic Sites." *New York Times,* p. A1.

———. 2003 May 27. "U.S. Report Faults Efforts to Track Water Pollution." *New York Times,* p. A1.

Sellars, Richard West. 1997. *Preserving Nature in the National Parks: A History.* New Haven: Yale University Press.

Sellers, Jefferey. 2002. *Governing from Below: Urban Regions and the Global Economy.* New York: Cambridge University Press.

Sexton, Patricia. 1991. *The War on Labor and the Left.* Boulder: Westview.

Shaiko, Ronald. 1999. *Voices and Echoes for the Environment.* New York: Columbia University Press.

Simonis, Udo E. 1989. "Ecological Modernization of Industrial Society: Three Strategic Elements." *International Social Science Journal* 41, no. 3: 347–361.

Skocpol, Theda. 1979. *States and Social Revolutions*. Cambridge, U.K.: Cambridge University Press.

———. 1985. "Bringing the State Back In: Strategies of Analysis in Current Research." In *Bringing the State Back In*. Edited by Peter Evans, Dietrich Rueschemeyer, and Theda Skocpol. Cambridge, U.K.: Cambridge University Press.

———. 1986/7. "A Brief Response [to G. William Domhoff]." *Politics and Society* 15, no. 3: 331–332.

———. 1992. *Protecting Soldiers and Mothers: The Political Origins of Social Policy in the United States*. Cambridge: Harvard University Press.

Skocpol, Theda, Marshall Ganz, and Ziad Munson. 2000. "A Nation of Organizers: The Institutional Origins of Civic Voluntarism in the United States." *American Political Science Review* 94, no. 3: 527–546.

Skowronek, Stephen. 1982. *Building a New American State: The Expansion of National Administrative Capacities, 1877–1920*. Cambridge, U.K.: Cambridge University Press.

Smith, Eric R. A. N. 2002. *Energy, the Environment, and Public Opinion*. Lanham, Md.: Rowman & Littlefield.

Smith, Michael Peter. 2001. *Transnational Urbanism: Locating Globalization*. Malden, Mass.: Blackwell.

"Smog City." 1999 Nov. 4. *Houston Chronicle*, p. YO7.

Snell, Bradford C. 1974. *American Ground Transport*. Washington, D.C.: U.S. Government Printing Office.

Sonenshein, Raphael. 1993. *Politics in Black and White*. Princeton: Princeton University Press.

Steele, Brian C.H. and Angelika Heinzel. 2001 . "Materials for Fuel-Cell Technologies." *Nature* 414 (November): 345–352.

Steinberg, Ted. 2002. *Down to Earth: Nature's Role in American History*. New York: Oxford University Press.

Stigler, George J. 1971. "The Theory of Economic Regulation." *Bell Journal of Economics and Management Science* 2 (spring): 3–21.

Stilgoe, John. 1988. *Borderland: Origins of the American Suburb, 1820–1939*. New Haven: Yale University Press.

Stone, Clarence N. 1989. *Regime Politics*. Lawrence: University Press of Kansas.

Stradling, David. 1999. *Smokestacks and Progressives: Environmentalists, Engineers, and Air Quality in America, 1881–1951*. Baltimore: Johns Hopkins University Press.

Stradling, David, and Joel A. Tarr. 1999. "Environmental Activism, Locomotive Smoke, and the Corporate Response." *Business History Review* 73, no. 4: 677–704.

Szasz, Andrew. 1994. *Ecopopulism*. Minneapolis: University of Minnesota Press.

Tarr, Joel A. 1996. *The Search for the Ultimate Sink: Urban Pollution in Historical Perspective*. Akron, Ohio: The University of Akron Press.

Tarrow, Sidney. 1994. *Power in Movement*. New York: Cambridge University Press.

Taylor, Bron R., ed. 1995. *Ecological Resistance Movements*. Albany: State University of New York Press.

Tesh, Sylvia. 2000. *Uncertain Hazards: Environmental Activists and Scientific Proof*. Ithaca, N.Y.: Cornell University Press.

"Text of Report and Conclusions of Smog Expert." 1947 Jan. 19. *Los Angeles Times*, p. 1.

Thurow, Lester C. 2001. *The Zero-Sum Society: Distribution and the Possibilities for Economic Change*. New York: Basic.

"'Times' Expert Offers Smog Plan." 1947 Jan. 19. *Los Angeles Times*, p. 1.

Trenberth, Kevin E. 2001. "Stronger Evidence of Human Influences on Climate." *Environment* 43, no. 4: 8–18.

United States Congress (U.S. Congress). 1990. *Hearings before the Subcommittee on Energy and Power of the Committee on Energy and Commerce*. U.S. House of Representatives, Oct 18–19 1989, No. 101–120. Washington, D.C.: U.S. Government Printing Office.

———. 1998. *The Kyoto Protocol and Its Economic Implications: Hearing Before the Subcommittee on Energy and Power of the Committee on Commerce, House of Representatives, One Hundred Fifth Congress, second session, March 4, 1998*. Washington, D.C.: U.S. Government Printing Office.

Useem, Michael. 1984. *The Inner Circle: Large Corporations and the Rise of Business Political Activity in the U.S. and U.K.* Oxford: Oxford University Press.

Uzawa, Hirofumi. 2003. *Economic Theory and Global Warming*. New York: Cambridge University Press.

Victor, David G. 2001. *The Collapse of the Kyoto Protocol and the Struggle to Slow Global Warming*. Princeton: Princeton University Press.

Viehe, Fred W. 1981. "Black Gold Suburbs: The Influence of the Extractive Industry on the Suburbanization of Los Angeles, 1890–1930." *Journal of Urban History* 8, no. 1: 3–26.

Vietor, Richard H. 1980. *Environmental Politics and the Coal Coalition*. College Station, Tex.: Texas A&M University Press.

Vos, Robert O. 1997. "Competing Approaches to Sustainability: Dimensions of Controversy." In *Flashpoints in Environmental Policymaking: Controversies in Achieving Sustainability*. Edited by Sheldon Kamieniecki, George A. Gonzalez, and Robert O. Vos. Albany: State University of New York Press.

Wachs, Martin. 1984. "Autos, Transit, and the Sprawl of Los Angeles: The 1920s." *Journal of the American Planning Association* 50, no. 3: 297–310.

Wainwright, Hilary. 1994. *Arguments for a New Left: Answering the Free-Market Right*. Cambridge: Blackwell.

Wald, Matthew L. 1989 April 7. "Alternative-Fuel Vehicles Move from Fancy to Fact." *New York Times*, p. A1.

Walker, Jack L. 1991. *Mobilizing Interest Groups in America*. Ann Arbor: University of Michigan Press.

Walker, Louise Drusilla. 1941. The Chicago Association of Commerce: Its History and Policies. Ph.D. diss., University of Chicago.

Wall, Derek. 1999. *Earth First! and the Anti-Roads Movement: Radical Environmentalism and Comparative Social Movements*. New York: Routledge.

Ward, David. 1964. "A Comparative Historical Geography of Streetcar Suburbs in Boston, Massachusetts and Leeds, England: 1850–1920." *Association of American Geographers Annals* 54 (December): 477–489.

Warner, Kee, and Harvey Molotch. 2000. *Building Rules: How Local Controls Shape Community Environments and Economies*. New York: Westview.

Warner, Sam Bass. 1978. *Streetcar Suburbs: The Process of Growth in Boston, 1870–1900*. Cambridge: Harvard University Press. Originally published in 1962.

Warrick, Joby. 1997 Oct. 20. "Greenhouse Gases Rose 3.4 percent in 1996." *Los Angeles Times*, p. A8.

Weale, Albert. 1992. *The New Politics of Pollution*. New York: Manchester University Press.

Weaver, John C. 1984. "'Tomorrow's Metropolis' Revisited: A Critical Assessment Urban Reform in Canada, 1890–1920." In *The Canadian City: Essays in Urban and Social History*. Edited by Gilbert A. Stelter and Alan F. J. Artibise. Ottawa: Carleton University Press.

Weber, Edward P. 1998. *Pluralism by the Rules: Conflict and Cooperation in Environmental Regulation*. Washington, D.C.: Georgetown University Press.

Weinstein, James. 1968. *The Corporate Ideal in the Liberal State, 1900–1918*. Boston: Beacon Press.

Weiss, Marc. 1987. *The Rise of the Community Builders: The American Real Estate Industry and Urban Land Planning*. New York: Columbia University Press.

Weisser, Victor, President, California Council for Environmental & Economic Balance. 2000. Interview by author, 13 March, San Francisco. Tape recording.

Weisskopf, Michael. 1990 Oct. 22. "Conferees Reach Acid Rain Accord." *Washington Post*, p. A1.

West, Darrell and Burdett A. Loomis. 1999. *The Sound of Money: How Political Interests Get What They Want*. New York: Norton.

White, V. John, Center for Energy Efficiency and Renewable Technologies. 2000. Interview by author, 14 March, Sacramento. Tape recording.

Whitt, J. Allen. 1982. *Urban Elites and Mass Transportation*. Princeton: Princeton University Press.

Widney, R. M. 1956. "Los Angeles County Subsidy." *The Historical Society of Southern California Quarterly* 38 (Dec.): 347–362. Originally published in 1872.

Wiewel, Wim, and Joseph J. Persky, eds. 2002. *Suburban Sprawl: Private Decisions and Public Policy.* Armonk, N.Y.: M. E. Sharpe.

Williams, Bruce, and Albert Matheny. 1984. "Testing Theories of Social Regulation: Hazardous Waste Regulations in the American States." *Journal of Politics* 46, no. 2: 428–59.

Williams, James C. 1997. *Energy and the Making of Modern California.* Akron, Ohio: University of Akron Press.

Wood, Dan B. 1988. "Principals, Bureaucrats, and Responsiveness in Clean Air Enforcements." *American Political Science Review* 82, no. 1: 213–234.

———. 1992. "Modeling Federal Implementation as a System: The Clean Air Case." *American Journal of Political Science* 36, no. 1: 40–67.

Wynne, Brian 1982. *Rationality and Ritual: The Windscale Inquiry and Nuclear Decisions in Britain.* Chalfont St. Giles, Bucks, U.K.: British Society for the History of Science.

Yago, Glenn. 1984. *The Decline of Transit: Urban Transportation in German and U.S. Cities, 1900–1970.* New York: Cambridge University Press.

Yeager, Peter C. 1991. *The Limits of Law: The Public Regulation of Private Pollution.* Cambridge, U.K.: Cambridge University Press.

York, Richard, and Eugene A. Rosa. 2003. "Key Challenges to Ecological Modernization Theory." *Organization & Environment* 16, no. 3: 273–288.

Young, Stephen C., ed. 2000. *The Emergence of Ecological Modernisation: Integrating the Environment and the Economy?* New York: Routledge.

Index

www.ingramcontent.com/pod-product-compliance
Lightning Source LLC
Chambersburg PA
CBHW020356270326
41926CB00007B/455